LONDON WATER HERITAGE
Professor S K Al Naib

Preface

London enjoys a rich and varied water inheritance from her historic past. Evidence can be traced of how Londoners have been shaping and altering their water environment for hundreds of years even before the coming of the Romans. It is this inheritance which has been commemorated in this book by Professor S K Al Naib, the ninth of his fascinating and successful books on London and Docklands.

Water is a commodity Londoners have taken for granted since the beginning of the 20th century. Few human needs have inspired such heroic feats as a reliable supply of fresh water and wastewater disposal. The River Thames winding through the great Capital supports a vast network of man-made water systems which sustain a high quality of life where it could not otherwise survive. Enjoy your heritage!

Content

	Page
London Subterranean World	3
Prehistoric Thameside	4
Thames Water Time Chart	6
Early London Water Supply	8
History of London New River	10
London Early Public Utilities	14
Explore London Canals	16
18th & 19th Century Water Companies	18
21st Century London Water	22
London Main Drainage	24
Discover the River Thames	30
New Docklands Waterside	32
Thames Wildlife	34
Thames Bridges Heritage Trail	35
London Safe from the Sea	38
Acknowledgements and Information	40

Copyright S K Al Naib

All rights reserved. No part of the book may be reproduced or transmitted by any means without prior permission of the copyright holder. Whilst every care has been taken to ensure the accuracy of this publication, the author and publisher cannot be held responsible for any errors. The views expressed are of the author and do not necessarily represent the opinions of the University of East London.

"Londons Water Heritage"   ISBN 1 8745 36 406
First Printing: January 2001

Internationally Acknowledged Books by Prof Naib

"Londons Water Heritage" 2000 years Heritage ISBN 1 8745 36 406
"London and Docklands Walks" The Explorer   ISBN 1 8745 36 252
"London Millennium Guide" Ed., Ent., & Asp.   ISBN 1 8745 36 201
"London Dockland Guide" Heritage Panorama   ISBN 1 8745 36 031
"London Illustrated" History, Current & Future ISBN 1 8745 36 015
"Discover London Docklands" A to Z Guide   ISBN 1 8745 36 007
"London Docklands" Past, Present and Future ISBN 1 8745 36 023
"European Docklands" Past, Present & Future ISBN 0 9019 87 824
"Dockland" Historical Survey of East London   ISBN 0 9089 87 800
"Fluid Mechanics, Hydraulics and Envir. Eng." ISBN 1 8745 36 066
"Applied Hydraulics, Hyd and Envir. Eng."   ISBN 1 8745 36 058
"Jet Mechanics and Hydraulic Structures"   ISBN 0 9019 87 832
"Experimental Fluid Mechs & Hyd Modelling"   ISBN 1 8745 36 090

See information on the UEL web site **http://www.uel.ac.uk**

The author is Professor of Civil Engineering and Head of Department at the University of East London, Longbridge Road, Dagenham, Essex, RM8 2AS, Great Britain.

(Tel: 020 8223 2478/2531,   Fax: 020 8223 2963)

> **PLEASE ORDER THROUGH:**
> **RESEARCH BOOKS, P O. BOX 82, ROMFORD, ESSEX, RM6 5BY, GREAT BRITAIN**

St Tropez? Puerto Banus? No, this is the Limehouse Basin Marina in London Docklands.

London Subterranean World

Subterranean London
London has an extensive subterranean infrastructure including crypts, tunnels, pipes, sewers and cables, some with little or no written documentation. Some of these features date back many centuries but most of them were built after the middle of the 19th century. The subterranean world of services such as gas, electricity, water, sewerage, cable TV, telecommunications and transport systems, remain a mystery to many people. In this section the history and development of mains services and infrastructure are briefly outlined.

London's Water Supply
Early water supplies were either from wells or rivers. The capital had been built on water bearing gravel beds and springs abounded. The Walbrook and Fleet rivers (now underground) brought supplies to the City from the hills to the north. During the 13th century, conduits were constructed to bring water from outlying springs to points around the City. General distribution was by means of water carriers known as cobs. In 1581 the first water wheels at London Bridge were used to pump water. Early in the 17th century water was carried along a canal, called the New River, which followed the Lee Valley and still operates today. By 1850, eight companies supplied Greater London through an extensive network of underground pipes. They merged in 1903 to form the Metropolitan Water Board.

In 1974 a major reorganisation took place when England and Wales were divided into ten major regions each under a Water Authority. Thames Water Authority took over the management of the whole of the Thames Valley from the Cotswolds to Gravesend and became responsible for water supply, drainage and sewage disposal. During 1992, the authority was privatised and now functions under various companies.

New London Water Ring Main
The new London ring main has recently been constructed as an 80 km tunnel from Ashford to Sunbury in West London to Coppermills at Walthamstow in the North East. The huge tunnels run 40 metres below ground and supply half of London with water. Substantial extensions are proposed during the early 21st century.

London's Main Sewers
Following the Public Health Act of 1858, Sir Joseph Bazalgette, the Chief Engineer of the Metropolitan Board of Works, devised a system for sewage disposal in London which is still in operation. Five main sewers, three north of the Thames and two to the south, carried the sewage over 10 miles to Barking and Plumstead. One of the conduits is part of the Victoria Embankment which Bazalgette also designed. Here 11 hectares of land was reclaimed from the Thames reducing its width and removing the mudflats which were mainly deposits of sewage discharged into the river at that time and resulted in bad smells in the area. The massive retaining wall stretched for one and a third miles from Westminster to Blackfriars in a gentle curve along the line of the Thames. The Victoria Embankment transformed the north bank of the river, relieving congestion in the Strand, and becoming a pleasant place for a stroll or rest with fine views and public gardens.

Underground Tube Lines
London had the world's first underground railway, the Metropolitan Line, which ran four miles from Paddington to Farringdon Street. Opened in 1883, it took three years to construct and was an instant success; within six months 26,000 people were using it daily. In order to relieve traffic congestion within the City of London, the underground railway was expanded to join the major stations. The Circle Line followed in 1884. The District and Metropolitan Lines expanded soon after from the Circle Line into the suburbs and helped considerably in new housing developments and the establishment of Greater London. This period also marked the introduction of electric trains to replace steam. Tunnelling techniques were also improved which resulted in more tubes. The City and South London was followed by the Waterloo and City in 1898, the Bank to Shepherd's Bush 1900, Baker Street and Waterloo 1906, Piccadilly and Brompton 1906, and Charing Cross, Euston and Hampstead 1907. No other tubes were built until the Victoria Line to Walthamstow in 1971, the Jubilee Line in 1979 and the Jubilee Line extension from Green Park to Stratford in 1999.

Gas and Electricity
In 1804 the New Light and Heat Company was established and lit the Lyceum Theatre with gas. By the 1840s most of London's main streets were gas-lit and twelve companies existed consuming nearly a quarter million tonnes of coal each year and employing about 400 lamp-lighters. Underground iron pipes were used for carrying the gas.

Electric lighting first appeared in London after the completion of Westminster Bridge in 1858, but the use of electricity did not spread widely until the beginning of the 20th century. Today the Capital is lit entirely by electricity and the romantic cast-iron gas lamps have disappeared forever.

Natural gas from the North Sea is used for cooking and heating with their pipes and the electricity cables buried under the pavement. London Electricity Company has an underground station under Leicester Square in the West End opened in 1992, with access through a pavement hatch.

Royal Mail Trains
The Royal Mail Company operates electric underground railways which run between mail sorting offices and British Rail main line stations. The trains are fully automated and driverless. One of these mail trains runs at 20-30 metres underground from Mount Pleasant sorting office and control room to Paddington.

Brunel's Thames Tunnel
The year 1993 was the 150th anniversary of the opening of Sir Marc Brunel and his son Isambard Brunel's Thames Tunnel. When opened in 1843 the pioneering tunnel had taken 18 years to complete. It was the first river tunnel in the world and was sold in 1865 to the East London Railway Company. Today it still carries the trains of the underground East London Line. There is an excellent exhibition at the Brunel Engine Museum in Rotherhithe.

Blackwall Tunnel
The 1.4km northbound bore of the Blackwall Tunnel was opened in 1897 and the southbound in 1967. The traffic flow for the first year of operation was almost 10,000 vehicles a day, mostly horse-drawn wagons. Currently the northbound traffic flow is over 40 times that figure.(Though not so many horse drawn vehicles!)

Dartford Tunnel
The first tunnel opened in 1963, forms the northbound crossing of the M25 motorway on the east side of London.

Chislehurst Caves & Mines
Mined by the Romans and Londoners for hundred of years, there are a large number of chalk caves linked by many kilometers of passages, several of which are open to the public. British Rail station is directly opposite the caves.

Prehistoric Thameside

Bronze and Iron Age Fisherfolk

The history of the eastern part of Docklands, The Royals, has been traced back some 3000 to 6000 years through archaeological finds. The Bronze and Iron Age inhabitants of this marshy Thameside were most probably fisherfolk and hunters. From the remains of the substantial timber track discovered, Museum of London archaeologists believe the area sustained a number of settlements. Later the Romans had a burial ground nearby.

Evidence also suggests there could have been a Roman road and a ferry point, and perhaps a look-out river post at Gallions Reach. During the 9th century, Bow Creek and Barking Creek were hideouts for Viking raiders, which patrolled the area during their attacks to England.

In Medieval Times, the area was known as Hamme which was ceded to the East Saxons by King Alfred the Great. Later King Offa gave the land to the Monastery of St Peter's at Westminster Abbey. Little is known about the district pre-1700 although cattle were grazing on the marshes. By the end of the 18th century, there was only one house in the marshland and a road stretching from East Ham village to the Thames.

Wild Animals in Docklands

A map of the City of London and surrounding villages of 1799 shows the north side of the Thames from the Isle of Dogs, through the Royal Docks to Aveley in Essex, as marsh and wetland areas, lying very close to the present day A13 trunk road. These were probably similar to Rainham marshes but with forests and woodlands. Indeed, excavations in the Thames Estuary at Aveley marshes near Purfleet during the 1990s uncovered an exciting range of fossils, including the remains of a large lion, brown bear, wolf, horse, red deer, rhinoceros, beaver, mole, bat, water vole and pond terrapin. The rich organic deposits also preserved remains of various birds, amphibians, beetles and shells. All of these creatures probably roamed the Royals marshes and fen land.

A major find was the bones of a jungle cat (Felis Chaus) about 40cm (16 inches) in height. These are now found only in the marshlands of India, South East Asia and North Africa and are often called Swamp Cats. Much larger bones were found from a giant ox (Bos Primigenius), a prehistoric auroch and possibly the forerunner of our domestic bull. They stood about two metres high at the shoulder and had long horns. They probably became extinct about 4000-5000 years ago, during the Bronze Age. The skeleton is now preserved in the Natural History Museum at South Kensington, London.

The fossils were uncovered during the construction work on the Wennington to Mardyke section of the A13, close to the M25 motorway. They were found in sediments which were deposited by the River Thames approximately 200,000 years ago, in the Middle Pleistocene period. The climate during this interglacial period was at least as warm as todays with similar environmental conditions. The area was of marshy vegetation close to the bank of the river, covered with extensive grasslands and dotted with patches of woodland. The Church used the grassland around Docklands for fattening their cattle during the 17th and 18th centuries.

In the 1960s the Natural History Museum excavated on the north side of Aveley village and exposed similar sediments, where skeletons of the woolly mammoth and of the straight-tusked elephant (Plalaeoloxodon Antiguus) were discovered. Earlier excavations north of Kew Bridge in the early 1800s yielded similar mammal remains in the Thames river drift deposit of sand and gravel 2 metres thick, implying that these animals probably wandered along the whole length of the Thames Valley.

6000-year Old Docklands Tree

The coming of the railway during the 19th century opened up the area with the construction of the East Counties Railway in 1847. There were a few cottages and the Old Barge House, from where ferries crossed to Woolwich. Later in 1855 Prince Albert opened the Royal Victoria Rock. During excavations for this dock hazel, oak and yew trees were discovered in a bog, part of a forest found in the area, estimated to be 6000 years old. Other finds included English and Roman coins, a millstone, a Roman urn, a tin circular shield and many animal bones similar to those found at Aveley marshes. An 8 metre dugout canoe was excavated and removed to the British Museum. Industrial activities and raising the land during the construction of the docks may have destroyed archaeological remains and sadly the potential for organised future excavation is therefore very limited.

An impression of prehistoric animals in Docklands.

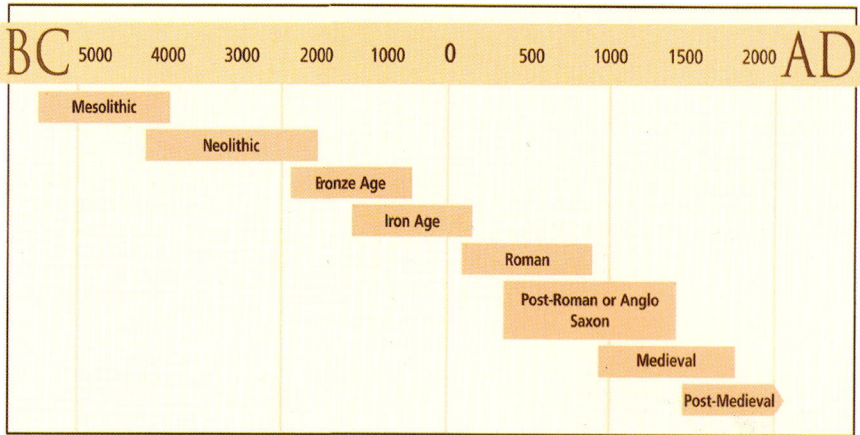

 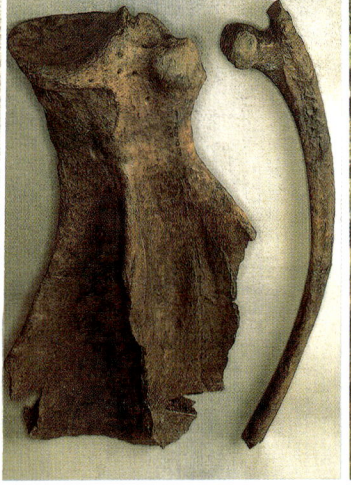 

Left: Foreleg of a jungle swamp cat. Top: Shoulder blade and rib of a narrow-nosed rhinoceros. Right: Typical Iron Age settlement of round houses with a village hall.

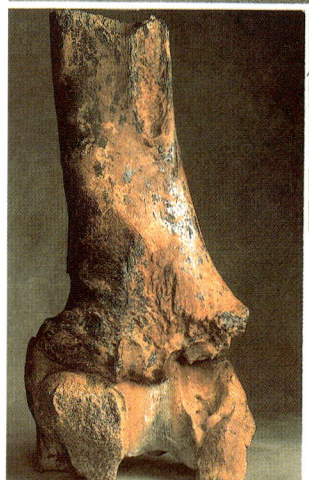 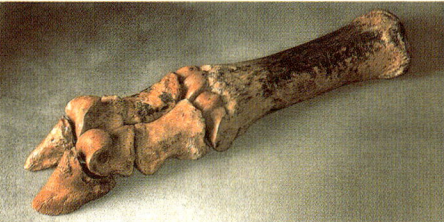 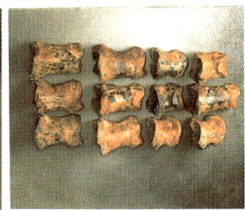

Top Left: Bones and hoof of the aurochs, 200,000 years old, discovered close to the A13 at Aveley.

Top: Toe bones from aurochs, the giant ox.

Right: Typical Neolithic Age flint knife about 6000 years old.

Map of London circa 1799, showing marshy Thameside in Docklands.

The Anglo-Saxon pitched roof house, circa 5th century.

Top: Fishermen at Blackwall Reach, circa 1801.
Left: Barking Creek circa 1845, developed as a major fishing port over many centuries.

Thames Water

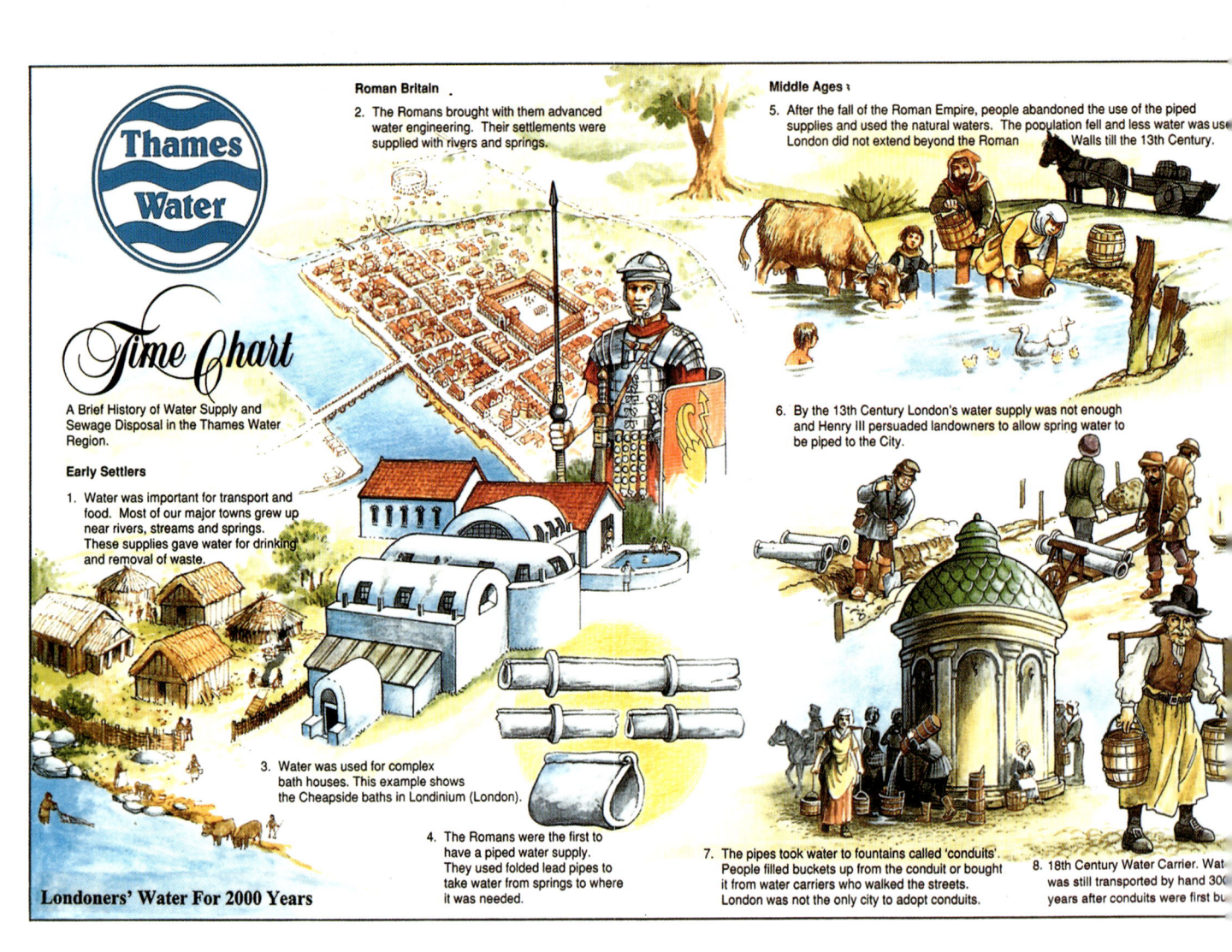

Time Chart

A Brief History of Water Supply and Sewage Disposal in the Thames Water Region.

Londoners' Water For 2000 Years

Early Settlers

1. Water was important for transport and food. Most of our major towns grew up near rivers, streams and springs. These supplies gave water for drinking and removal of waste.

Roman Britain

2. The Romans brought with them advanced water engineering. Their settlements were supplied with rivers and springs.

3. Water was used for complex bath houses. This example shows the Cheapside baths in Londinium (London).

4. The Romans were the first to have a piped water supply. They used folded lead pipes to take water from springs to where it was needed.

Middle Ages

5. After the fall of the Roman Empire, people abandoned the use of the piped supplies and used the natural waters. The population fell and less water was used. London did not extend beyond the Roman Walls till the 13th Century.

6. By the 13th Century London's water supply was not enough and Henry III persuaded landowners to allow spring water to be piped to the City.

7. The pipes took water to fountains called 'conduits'. People filled buckets up from the conduit or bought it from water carriers who walked the streets. London was not the only city to adopt conduits.

8. 18th Century Water Carrier. Water was still transported by hand 300 years after conduits were first bu[ilt]

19th Century

18. Big changes took place in water supply and sewage treatment, prompted by rising populations and the realisation that polluted water could cause disease.

19. A number of water companies sprang up across the region, charging their customers for the service. Water was usually taken from rivers and not treated in any way. In some cases it was stored in reservoirs before reaching the houses.

20. From 1830s - 1860s there were outbreaks of cholera across the country. In 1854 John Snow, a doctor in London, was the first to show that deaths were linked to cholera polluted water by showing that a cluster of deaths had occurred round one particular water pump near Regent Street.

21. In the 1850s Punch magazine, and others, campaigned for a clean water supply. Cartoons showed the drinking water to be filled with strange creatures, others brought attention to the filthy state of the River Thames. One water company took drinking water from the river directly opposite a main sewer!

22. The Metropolis Water Act of 1852 finally brought far reaching changes into being, to protect the water supply. Mainly centred on London, the principles were applied across the region and still apply today. They were: i) No river water to be taken downstream of Teddington; ii) All river water to be filtered; iii) Reservoirs for filtered water to be covered.

Areas served by the Metropolitan Water Companies, 1899

Punch's view of a drop of drinking water

23. The supply was helped by more efficient steam engines and the use of cast iron pipes which withstood higher pressures and did not leak as much. In 1871 water supply on a Sunday was made compulsory.

24. From the 1830s river water began to be filtered through beds of sand to remove particles from the water. Bacteriological examinations were not routine before 1885.

Grand Junction Waterworks, 1827

Claremont Square Reservoir, 1855

25. Most waste continued to find its way to the rivers or cesspits which were emptied by hand or small pumps. The increasing use of the flush toilet meant that much more water was used for the flushing. Many cesspits could not cope and overflowed to rivers.

26. The River Thames was so polluted by the 1850s that the Houses of Parliament could not operate properly. 1858 was known as the 'Year of the Great Stink' and led to changes in the law regarding sewage disposal.

Night soil men

Water Closet, 1890

Faraday giving his card to Father Tha[mes]

Illustrated by Ray Mutimer. Produced by MBA Publishing Ltd. 0937 - 844515

Early London Water Supply

Roman Period

The Romans developed water supply and sanitation systems in their towns and cities over two thousand years ago. Rome was initially dependent on water drawn from the River Tiber and private wells within the City. Population growth and the increasing sophistication of the Roman lifestyle, with its fondness for public baths and fountains, made it necessary for the Romans to substantially increase their supplies. Many aqueducts were built to bring water from outside the city to supply more than thirteen hundred public fountains, conduits and six large public baths, besides providing a piped water supply for the homes of the wealthy.

During their stay in Britain, 43-410AD, the Romans contributed to the development of water supply and drainage in forts, towns and villas such as Housesteads Fort on the Hadrian's Wall, Lincoln, York, Cirencester, and Dorchester. They also invented the force pump to assist in raising of water and the remains of one were found at Cirencester. In the City of London, local wells and rivers were sufficient to provide ample water supply. On the whole, the water engineering achievements of the Romans in Britain were not so spectacular as elsewhere in their European Empire. (see illustration opposite).

Early History

At the time of the Roman occupation London was a well watered city with the Thames as its southern boundary and the River Walbrook dividing it roughly into two halves. During the Norman period, John Stow wrote of the excellent wells and springs in the northern suburbs with their clear and sweet waters.

During the early 13th century buildings expanded outside the City and the population increased forcing new developments to seek clean water outside the city boundary. Work began to convey water from the River Tyburn by lead pipes into the city. In 1245 a conduit house was built in Cheapside. The pipes ran from the source to James Head, to Musegate then to the Cross in Cheap. James Head seems to have been on the site of the churchyard of St James, Piccadilly, and the term 'head' implies a spring supplying the water. It also appears that Musegate is the site of the Royal Mews in Trafalgar Square. The total distance was about 3000 metres.

This was the first of the London conduit supplies and others soon followed. There were no pumps and the water was fed under gravity due to the natural pressurehead. Distribution pipes were rather small hence Shakespeare's reference to the 'missing conduit' in Henry VI Part 2. The water was freely available to the citizens but trade users were assessed and charged. Keepers were appointed to manage conduits which were financed by rates levied on houses in the vicinity, by legacies and benefactors. Private supplies to houses were permitted only on special authority. At the coronation of Edward I in 1273, the Cheapside conduit flowed all day with red and white wine!

The quality of the river water was a matter of constant debate and concern. In 1297 the Earl of Lincoln complained to Parliament that river water was unfit for drinking due to the filth running into it from tanneries along its bank. In 1337 trade users were also accused, particularly the Breweries near the conduit, about the amount of wastewater which was discharged into the rivers.

By the beginning of the 15th century it became necessary to supplement the Tyburn waters to the City and a cistern was erected in Cornhill. In 1423 two additions to the supplies were made at the expense of Sir Richard Whittington. Springs in the Manor of Paddington were granted to the City in 1440 by the Abbot of Westminster.

Elizabethan Times

The first pumped water supply for London began in 1581 when the City Corporation granted Peter Morice a 500-year lease at 10/- (50p) annually on the first arch of London Bridge where a water wheel was constructed. Pumped supplies gradually replaced the gravity conduits. The first major mechanical scheme comprising a chain pump worked by horses supplied a tower about 120ft. high which was established by a mining engineer, Bevis Bulmer, at London Bridge. This served the Cheapside and St Paul's area. In about 1600 a petition was presented to the House of Commons from the whole company of poor water tankard bearers of the City of London and the suburbs, they and their families being about 4000 in number. It complained of legal and illegal depletion of the conduits supplies by private contractors. The trade of the water carriers had also been affected by the new pumped supplies.

Having so forcibly demonstrated the success of his system, Morice, supported financially by the Fishmongers Company, was able to install further water wheels under the arches of London Bridge and thus supply the eastern end of the City with water. These waterworks remained in place until 1822.

Water Bearers

During Elizabethan times people were mainly compelled to fetch water from cisterns or buy it from the Water Bearers. However, the rich merchants and certain influential people who prevailed upon the City of London Corporation were allowed to tap the conduits and run a private water supply to their houses using leaden pipes.

The Water Bearers had their own guild called 'The Brotherhood of Saint Christopher of the Water Bearers'. This association continued for 60 years after which the Water Bearers came under the control of the City Chamberlain. At the beginning of the 17th century Parliament authorised the construction of the New River to bring water to London from Chadwell and Amwell, in Hertfordshire.

Pipe Borers

The New River Company, having completed its channel, began distributing water around 1613 and for this purpose wooden water mains were used. Their use in Britain dates back to the Roman times. Sections dating from the occupation period have been found at Cirencester and Carement in Wiltshire. Iron collars for making connections were also discovered. In London the pipe-borer's work had become an important trade by the middle of the 17th century. John Evelyn, in his "Sylva" of 1670, showed one at work, boring out a tree-trunk. These trunks were usually of elm, though softer wood like fir was sometimes used. They were placed on a trestle and an augar, fitted to a large wooden handle, which resembled a giant gimlet, was used. This tool was turned by hand against the end of the trunk. The diameter of the bore ranged between 3 inches (75mm) and 7 inches (175mm). Later the New River Company had a horse-drawn mill at Dorset Stairs along the Thames, south of Fleet Street, which speeded up the boring process substantially and reduced the number of pipe borers employed. The wooden pipes lasted about 5 years but those buried in clay could last up to 25 years.

It was not until 1785 that Thomas Simpson invented the pipe spigot and fausett joint which provided a satisfactory method of connecting iron pipes. Around 1817 the wooden pipes were being replaced by cast-iron mains. Sections of wooden water mains were unearthed at Buckingham Palace, Westminster, in 1947 and were thought to be installed by the New River Company.

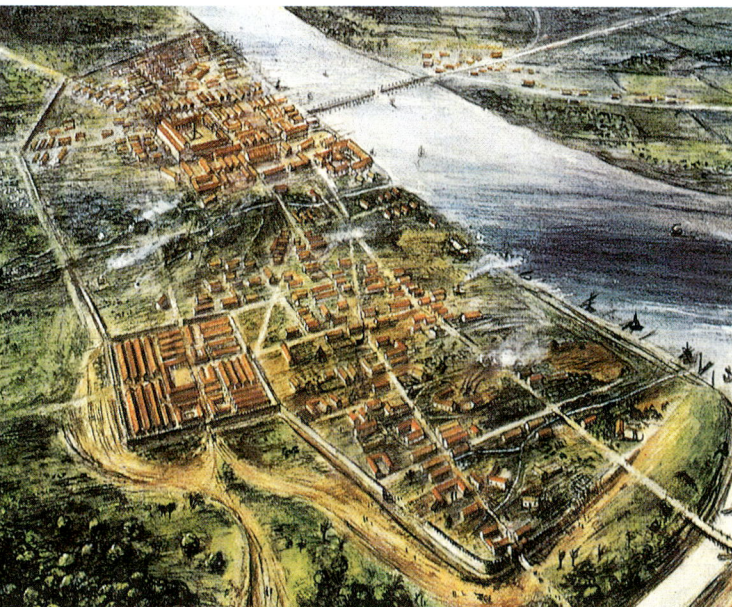

Top: A bird's eye view of the City of London looking south-east during Roman times, served by Walbrooke River at its centre and by the River Fleet on the west side.
Left: Aerial view in 1999 of the Roman aqueduct built in the Spanish town of Segovia in the first century AD.

An elevation of the aqueduct showing the spectacular legacy of Roman engineering in the town of Segovia, 70km north-west of Madrid.

An elevated view of Roman London looking north-west showing the timber London Bridge and the second century Basilica.

The Waterworks of London Bridge for the supply of the City of London with Thames water. This diagram shows one of the waterwheels with lifting apparatus, from the Universal, May 1794. The pumps were first installed around 1570-80 and were mounted in the first and second arches of the old stone London Bridge. The weir effect either side of the bridge was about 6ft.

History of London New River

Building the New River

Until the early 17th century, the water supply for London was limited to the Thames, its tributaries, wells, springs and conduits many of which were becoming contaminated. The idea of building a new river to resolve the problem was first suggested in 1580. Nothing happened however, until the turn of the century when Captain Edmund Colthurst of Bath suggested a scheme to bring water from springs in Hertfordshire and Middlesex to London. With the death of Queen Elizabeth I the application of letters patent was not considered. They were however granted in 1604 by King James I.

The proposals then went before the common Council of the City of London and a Bill was passed by Parliament in 1606 for the bringing in a fresh stream of running water to the northern parts of the City of London. Colthurst tried to win the contract to build the new river but in the end it was awarded to Hugh Myddelton a London goldsmith in 1609. Myddelton agreed to build a new river within four years, however the project was opposed by landowners who believed their farms and meadows would be destroyed. In May of 1609 the year work on the river which had then reached Cheshunt was stopped by an Act of Parliament. With money running out Myddelton approached James I who agreed to provide half the cost of the work in return for half the profits. Not surprisingly given the backing of the King, work was resumed in the autumn of 1611.

The project was completed on time, the London end of the New River - a pond just below Sadlers Wells known as the ducking pond, was reached in April 1613. It was later called the Round Pond and soon became known as the New River Head. It was situated roughly 80 ft above the level of the Thames and it was an ideal position for a reservoir from which water could be piped to houses in the surrounding area. Work on the pond and on the river itself continued until September 1613. Then finally on 29th September a formal inauguration ceremony took place and a celebration was staged for the completion of the New River.

Running for over 60 km, the river fell just 6 metres between its source at Chadwell Springs in Hertfordshire and its head, the Round Pond just below the present site of Sadlers Wells in Islington an average of just 78mm each kilometre. It was built with a minimum of surveying equipment by two hundred labourers and a number of the most skilled craftsmen of that period. At New River Head a cistern was soon built to the south of the Round Pond together with a building called The Water House, which was part office, part control station and part domestic accommodation for the supervisors for the New River Company. The water house contained the Oak Room, an elegant construction carved by Grindling Gibbons which later became part of the New River Head offices of the Metropolitan Water Board, Thames Water predecessors. These have recently been converted into private apartments.

In 1709 pumping of water from New River Head to a pond on the site of Claremont Square reservoir on Pentonville Road began. This was carried out between 1709 and 1720 by a windmill, later to be replaced by horsepower and then in 1768 by a steam pumping engine, the first to be used to pump drinking water to a city. The Round Pond was drained in 1913 and a year later the water house was demolished. However, many of the historic features remain including the old pond walls, the stump of the windmill, the pump house, the Metropolitan Water Board headquarters - built in 1916 - and the laboratory built in the 1930s.

Four Centuries of Service

Soon after the completion of the New River the supply from the springs was supplemented by water drawn from the River Lee near Hertford. An act of 1739 provided for the control of the quantity of water passing from the Lee into the New River controlled by means of a balance engine and a gauge, later known as the Marwell Gauge. The River Lee Act of 1855 gave to the New River Company and the East London Water Works Company the right to take the whole of the water in the River Lee with the exception of the quantity required for navigation. The New River Company was entitled to the first 22½ million gallons a day. The Act also substituted another gauge known as the New Gauge for the old balance engine and Marble Gauge which still exist today.

In 1833 two reservoirs were constructed along the New River at Stoke Newington in north London to be used partly as a reserve and partly to purify the water by allowing suspended matter to settle. As a result of the Metropolis Water Act of 1852 the Company built filtration works at New River Head and at Stoke Newington in 1856 and at Hornsey in 1859. During the 19th century the length of the New River was reduced to about 43 kilometres by straightening the course of the channel. The most important example of this work was completed in 1859 between Palmers Green and Hornsey High Street where a large double loop was dispensed with by cutting a new channel and a tunnel two-thirds of a mile long through the intervening pond.

Metropolitan Water Board

When the company was taken over by the Metropolitan Water Board in 1904 it was supplying about 1¼ million people with an average daily quantity of 41 million gallons. Between one-third and one-half of the supply was derived from wells. The Round Pond at New River Head was abandoned in 1914 and the site was later occupied by the headquarters of Thames Water Authority created in 1974. In 1946 the last of the filter beds at New River Head were taken out of use and the river ceased to flow there. The New River which continues to supply London with water now ends at the works in Green Lanes, Stoke Newington, and its length is 39 kilometres. The original source, the Chadwell spring is still used. A second source, the Amwell spring, has dried up but since 1739 water has been taken from the River Lee above Ware in Hertfordshire to feed the New River.

The illustrations on the opposite page show a heritage trail of the river in its heydays of 1888. Today, a section of it is preserved in Islington as ornamental gardens (see a and b on map, p12).

Sadlers Wells

In 1688 the New River Head witnessed the opening of nearby Sadlers Wells which started life as a centre of healing and music on two wells discovered by Richard Sadler. Unfortunately the theatre developed a reputation for rowdy behaviour and the worst example of this was a brawl in 1807 when eighteen people were killed. By the 1830s it was virtually bankrupt. This was changed under the management of actor Samuel Phelps who produced Shakespeare plays between 1844 and 1862 and completely changed the theatre's reputation. The cycle of success and failure continued until the last rebuilding in 1931 and the theatre's transformation by Lillian Bayliss. As a result of efforts by a number of people the theatre's vital role on the development of British opera and ballet was established, a role which it still enjoys today in their new building.

1. New River Source, Chadwell Springs, Ware.
2. New Gauge, King's Mead, Hertford.
3. Foot Bridge, Amwell Pond, Herts.
4. Turnford Pumping-Station, Cheshunt.
5. Theobalds Park.
6. Sluice House, Enfield.
7. Hornsey Pumping-Station.
8. Stone at Chadwell Springs, inscribed with date of New River works.
9. Pumping-Station, Green Lanes, Stoke Newington.
10. Engine-room of Pumping-Station.
11. New River Head, Clerkenwell.

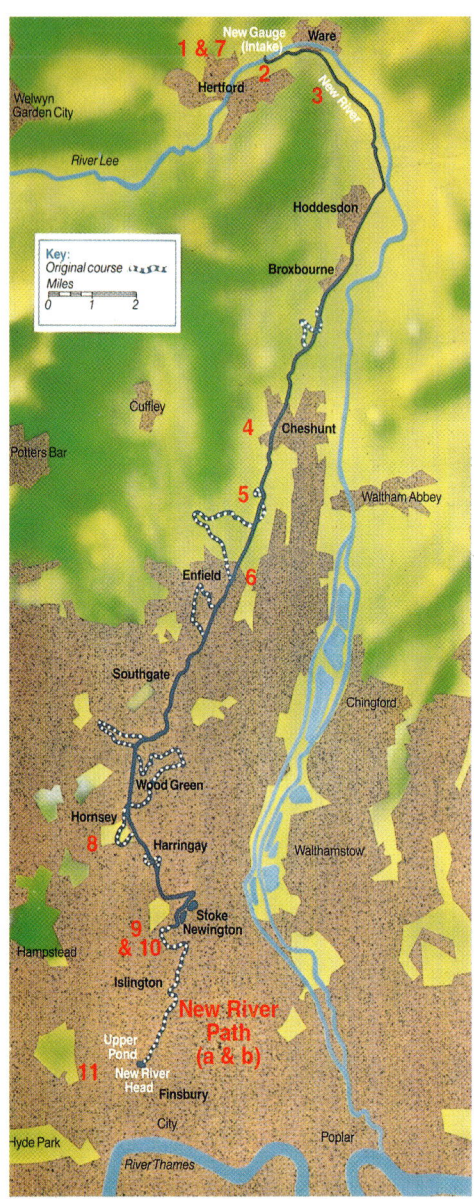

The current map showing the New River and River Lee.

Portrait of King James I, who put up half the money towards the cost of building the New River between 1609 and 1613.

Sir Hugh Myddelton, who built the scheme, was a Hatton Garden Jeweller.

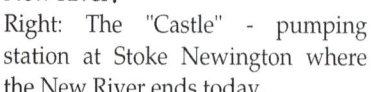

Top: The New Gauge, intake for the New River.
Right: The "Castle" - pumping station at Stoke Newington where the New River ends today.

Top: "Sir Hugh Myddelton Glory", an engraving of the opening ceremony in 1613 when water was let into the Round Pond at Islington. Myddelton is standing next to his horse on the right.
Right: A picture of the 400-year old New River, which during the new millennium is a key to the recharge of chalk aquifer in North London by pumping water into it from the river. The insets show the plan for the scheme and one of the new pumping stations, c1999.

The magnificent Oak Room was created during the 17th century on the first floor of the New River Head for the Company's committee meetings. Grinling Gibbons was responsible for much of the wood carving.

Details of Gibbons' carvings which reflect the abundance of nature - fruit, flowers, birds and fish.

Grinling Gibbons

The Oak Room today showing the wooden carved angels and the Royal Arms carved above the fireplace, part of the tribute to the King's success in ending the war with France in 1670s.

Henry Cooke was commissioned to paint the portrait of the reigning monarch, William III, on the ceiling.

Panel carving detail.

Statue of Sir Hugh Myddelton at Islington Green.

All that remains of the 18th century windmill is the sturrys at the New River Head today.

Top Left: The New River Head today, converted into private apartments.
Bottom Left: The New River ends at Green Lanes, Stoke Newington today.
Top Right: A section of the New River preserved as ornamental gardens in Islington after it ceased to flow to the New River Head.

13

London Early Public Utilities

Parish Utilities

Before the Paving Acts of 1761, householders were responsible for paving and lighting the streets outside their properties. This was often neglected and paving levels were uneven and dangerous. Under the Acts, Boards of Commissioners for each parish were made responsible for over-seeing the paving of streets with granite blocks laid in a bed of compacted soil. As the roads were improved, the central drainage gutters were replaced with narrower ones at the edges of the roads. Street lighting, prior to 1761, was generally poor. 'Link-men' and boys, carrying pitch torches, guided people through the streets. In richer areas householders provided whale-oil lamps outside their homes. By 1780, substantial progress had been made and it was said that there were more lamps in Oxford Street than in the whole of Paris!

Water for fire fighting was mainly provided by wooden fire engines. Parish fire brigades were unreliable and those who could afford it subscribed to the increasingly popular insurance companies during the 18th century. The firemen were usually the Thames watermen who wore a distinctive livery and badge. Fire marks were attached to buildings to show which company the owner or tenant was insured with.

Transportation

Roads leading from London were poorly maintained and dangerous. It was reported in 1720 that in one week all the stage coaches coming into the city from Surrey were robbed! Later in the century Turnpike Trusts started to undertake the improvement and maintenance of major roads out of the Capital; tolls were collected at main exit points. From 1780 the Royal Mail coaches travelled the turnpikes free of charge and provided an efficient postal service. Thousands of licensed watermen carried passengers across the Thames in their ferries. The river was a major thoroughfare, despite the opening of Westminster and Blackfriars bridges in 1750 and 1769 respectively. However, moving around on foot had its hazards as the streets were littered with rubbish, animals, traders' barrows and thieves! As the City expanded, the use of horse-drawn transport increased. Hackney carriages for two or three passengers were popular within the City. Short journeys outside the City were served by coaches. A light covered carriage was developed for quick journeys within the City.

CROSSING A DIRTY STREET.

Crossing the poorly-kept streets was a hazardous business, as this early 18th century satirical engraving shows.

Grand Union & Regents Canals

At the beginning of the 19th century, water transport became the new and most efficient method of transporting goods, coal and other materials into the Capital. A network of canals were built by civil engineers which joined the Thames with the rivers Trent, Severn and Mersey. The great Port of London had thus direct water transport links with many of the country's major manufacturing and trading cities. Today, these canals are used by many pleasure cruises and the foot paths are used by walkers. Maps and information are given in the following pages on the Grand Union and Regents Canals which are among the most important and interesting pieces of industrial heritage in Great Britain.

Camden Little Venice

Little Venice is one of the prettiest and romantic waterside districts in London, a unique combination of white stucco houses, greenery and canal water. It contains one example of the finest early domestic architecture in the capital. Past famous residents include Sigmund Freud, Robert Browning, Edward Fox and Joan Collins. It is one of the loveliest canal-side streets in England with the Regent's Canal gliding gracefully past it. A cruise on the Regent's Canal to London Zoo or a walk along the towpath is studded with fragments of history bringing the history of London and the age of canals to life. It captures the Victorian era when London was the great cultural and commercial capital of the world.

Paddington Packet Boat was operated as a fast passenger vessel on the Grand Union Canal, early 1800s.

Regent's Canal waterbus passenger service cruises from Little Venice to London Zoo.

A view of the Basin of the Grand Junction Canal at Paddington, c1802.

Waterbus at Paddington Basin, Little Venice, c1990.

The Regents Canal eastern entrance to the Islington Tunnel, c1823.

Royal Cricketers Pub on the Regents Canal, c1990.

Fishing at the New River Head in Islington, late 18th century.

Hampstead Road Lockhouse at Camden, on the Regents Canal.

Explore London's Canals

Explore London's canals
Half-hidden in the bustling, urban setting of Britain's capital is the more secret, peaceful world of canals. The Grand Union, Regent's and Hertford Union Canals swing in a wide arc through the heart of London linking with the River Lee Navigation to the east and the River Thames to the south. It's a fascinating world to explore, weaving between factories and warehouses or opening out into parks, gardens and nature reserves and providing unusual views of some of London's best-known landmarks.

Growth, decline and revival
London's canals were not always so peaceful. In their heyday, they were a centre of industrial activity – providing an essential transport route from the River Thames to other parts of London, to Birmingham and the north. When these canals were built (the Regent's Canal was completed in 1820), they ran through open country, avoiding built-up areas. Wharves, warehouses and docks sprang up, turning the canalside into a lively commercial centre and inland port, allowing London to expand its trade. Cargoes included coal, timber, sand and gravel, metals, corn and hay, plus the more unusual items such as chocolate for Cadbury's factory and ice imported from Norway.

However, the canals soon faced fierce new competition from the expansion of railways. Only twenty-five years after the Regent's Canal was opened, its owners tried unsuccessfully to turn it into a railway line! For a short period the two transport systems worked together, with canal barges unloading goods on to trains. Remnants of this era can still be seen today where short canal extensions were built to reach right into railway depots. Gradually, the canals declined as trade was lost first to the railways and then to the roads.

When transport was nationalised in 1947 ownership of the canals passed first to the British Transport Commission and then in 1963 to the newly formed British Waterways Board.

Today, canals have developed an equally important role as a vital resource for recreation in the capital. They are also valued for their architecture and industrial archaeology and as an important refuge for wildlife.

Youth club and community boat, Pirate Castle, Camden

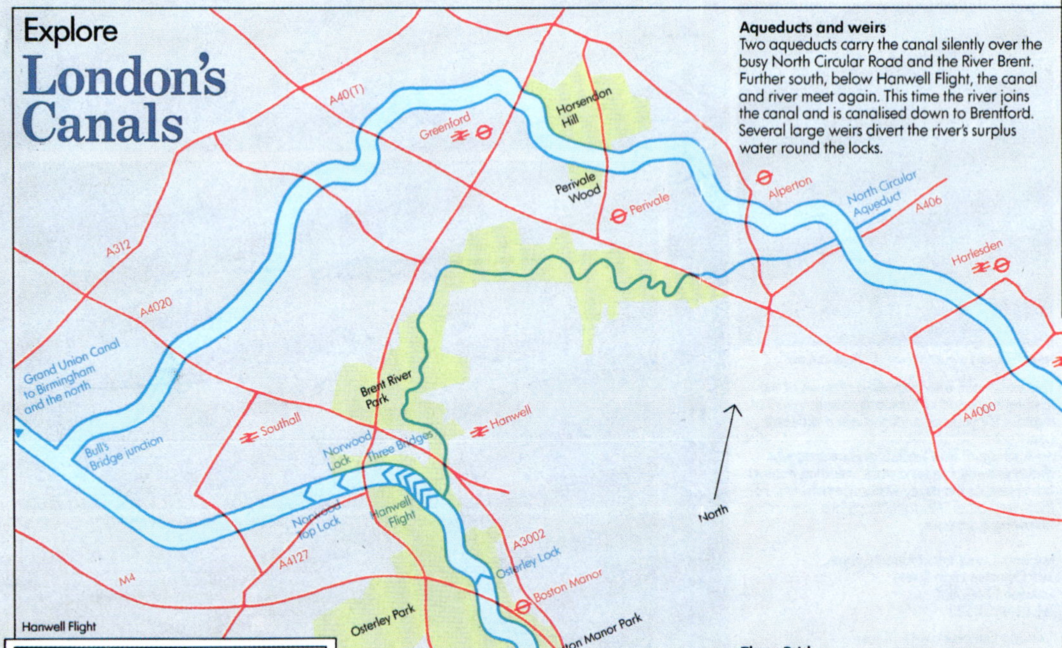

Explore London's Canals

Aqueducts and weirs
Two aqueducts carry the canal silently over the busy North Circular Road and the River Brent. Further south, below Hanwell Flight, the canal and river meet again. This time the river joins the canal and is canalised down to Brentford. Several large weirs divert the river's surplus water round the locks.

Three Bridges
A feature of canal architecture unique in London is the Three Bridges structure, a scheduled ancient monument designed in 1859 by the famous civil engineer Isambard Kingdom Brunel. Rail, canal and road pass one on top of another forming a fascinating intersection of transport routes.

Hanwell Flight
The Hanwell Flight of six locks raises the canal just over fifty-three feet in a third of a mile. The locks, early nineteenth century lock cottages and rural surroundings form an attractive setting. Take a walk along the Hanwell Flight and then follow the canal down through the wooded valley of the River Brent.

Trip boat, Regent's Canal

Little Venice

Royal Cricketer's pub, Old Ford Lock, Regent's Canal

London Canal Museum, Camden.

Walking
Leave the crowded London streets and take a quiet stroll beside the canal. Apart from short stretches such as the tunnels at Islington and Maida Hill and the Limehouse Cut, it is possible to walk along most of London's fifty-four miles of towing paths.

Many access points – all close to local bus routes or underground stations – make the canals ideal for short strolls or long-distance rambles. Whichever you choose, you will be rewarded by a rich variety of rural and urban scenery, glimpses of the canal's history and views of colourful narrowboats, locks and lock cottages. There is an abundance of plants and waterfowl too – in a setting not only ideal for walking, but also for photography, sketching, jogging or just relaxing in the numerous canalside pubs and restaurants.

In central London, access gates to the towing path are locked overnight. A booklet giving detailed information about canal walks in London is available from the Regent's Canal Information Centre, which can also provide details of guided walks.

Boating
There is plenty of opportunity for a variety of boating on London's canals. Particularly popular are the regular boat trips through Regent's Park between Little Venice and Camden, or trips from Camden along the Regent's Canal to the junction with the River Thames at Limehouse. The passenger boats also provide a novel way of reaching London Zoo.

Cruising restaurant boats provide an unusual setting for Sunday lunch or evening meals. For group outings or holidays, boats – including BWB's "Lady Rose of Regents" – are available for private charter or hire. Schools and community groups are also well served with canal boats available for day, weekend or week long holidays.

For those planning to visit London in their own boats there are a number of moorings, water points and other boating facilities available. Remember that all boats must have a BWB Pleasure Boat Licence.

Paddington Arm, Kensal Green

Canoeing
In the inner-city areas of London where facilities for outdoor watersports are limited, canals provide an ideal location for canoeing. There are a number of canoe clubs and canalside youth groups, open to both residents and visitors to London, providing facilities for canoeing, rowing and other water activities.

You can also use your own canoe after obtaining a BWB Pleasure Boat Licence or British Canoe Union membership. For safety reasons unpowered craft are not allowed in Maida Hill and Islington Tunnels, but the rest of London's canals are open to you – excellent for training, or simply for an enjoyable paddle. The twenty-two mile, lock-free stretch which runs from Camden along the Grand Union Canal west of London is particularly good for canoe touring.

Angling
Water quality in London's canals has improved so dramatically during the past decade that anglers quite regularly catch quality roach, bream, gudgeon, carp and tench in good numbers.

On the Regent's Canal restocking with fish netted from lakes in the Royal London Parks has significantly improved the fishing over recent years, particularly in Camden.

In common with the position elsewhere, you need to belong to the controlling club or purchase a day ticket from the patrolling bailiffs in order to fish. All anglers must also possess a Thames Water Authority rod licence before starting to fish.

Wildlife
One of the great values of London's canals is that they bring greenery and wildlife into the heart of urban and industrial areas of the capital.

Canals provide a wide range of habitats, with aquatic plants and vegetation along the waterside, and wildflowers, trees and shrubs beside the towing path. Swans, Canada geese, moorhens and ducks are common residents, with many other birds such as swallows, kingfishers and herons being less frequent visitors. You may even catch a glimpse of a newt or a watervole.

Although every care has been taken to ensure the accuracy of the information contained in this leaflet, BWB cannot accept responsibility for any error or omission.

Photographs from the British Waterways Board Photo Library. Designed by Sue Lamble. Printed by Fulmar Colour Printing Co. Ltd. Published by British Waterways Board 1986.

Front cover: St Pancras Lock

British Waterways Leisure

Little Venice
Little Venice is a triangular pool of water, created by the junction of the Paddington Arm, the Regent's Canal and the short arm leading to Paddington Basin. Colourful narrowboats and barges moor along the banks. Browning's Island, named after the poet Robert Browning, stands in the centre of Little Venice forming a haven for ducks, geese and swans beneath its weeping willows.

Little Venice is an excellent setting for a range of canal activities – canalboat trips to Camden and London Zoo, a canalboat art gallery, canalside pub, garden and walks.

London Zoo, Regent's Park

Regent's Park
The Regent's Canal sweeps through the northern section of Regent's Park and London Zoo in a deep tree-lined cutting, providing one of the most spectacular views on London's canals. On one side, the famous Snowdon Aviary rises above the canal, and the antelope terraces can be seen on the other.

Camden Lock Centre
Once a timber wharf and dock, this area has now been converted into a thriving craft and canal centre with craft shops, restaurants, boat trips, cruising restaurant boat and a bustling weekend market. The lock cottage has been converted to house BWB's Regent's Canal Information Centre.

Camden

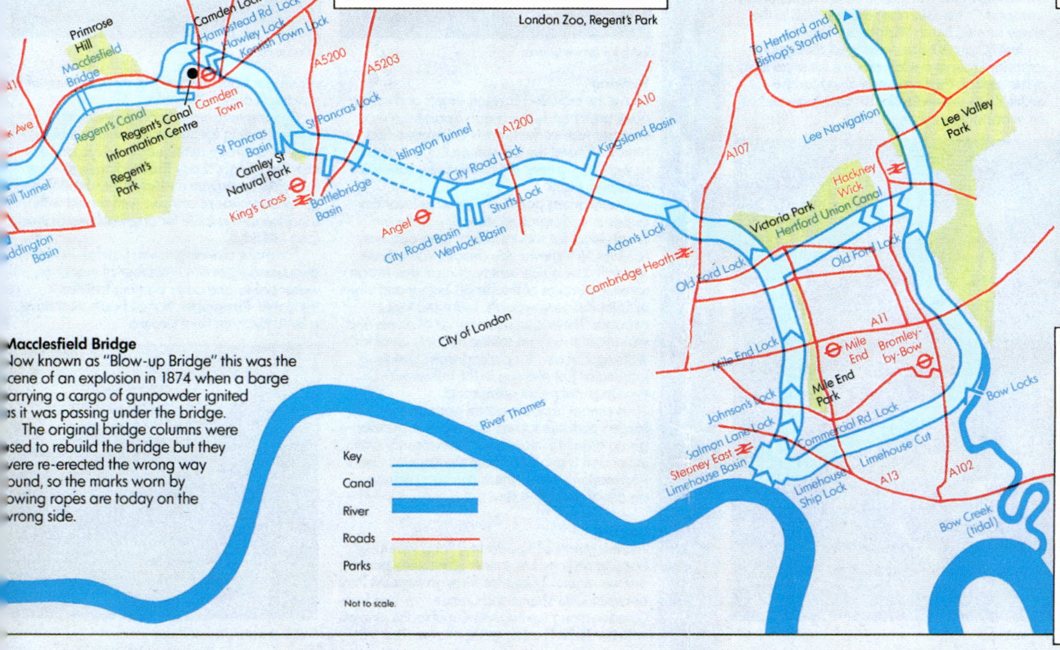

St Pancras Lock
In dramatic contrast to its urban surroundings the canal creates a picturesque waterside setting as it passes behind the busy King's Cross and St Pancras railway stations. St Pancras Lock and lock cottage and the moored boats in St Pancras Basin are complemented by the Camley Street Natural Park, an urban reclamation site next to the canal.

Limehouse ship lock
The Regent's Canal enters the River Thames through the largest canal lock in London. Ships from all parts of the world once entered the Regent's Canal through this ship lock to unload their cargoes onto canal barges and narrowboats.

St Pancras Lock

Macclesfield Bridge
Now known as "Blow-up Bridge" this was the scene of an explosion in 1874 when a barge carrying a cargo of gunpowder ignited as it was passing under the bridge.

The original bridge columns were used to rebuild the bridge but they were re-erected the wrong way round, so the marks worn by towing ropes are today on the wrong side.

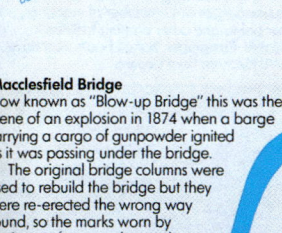

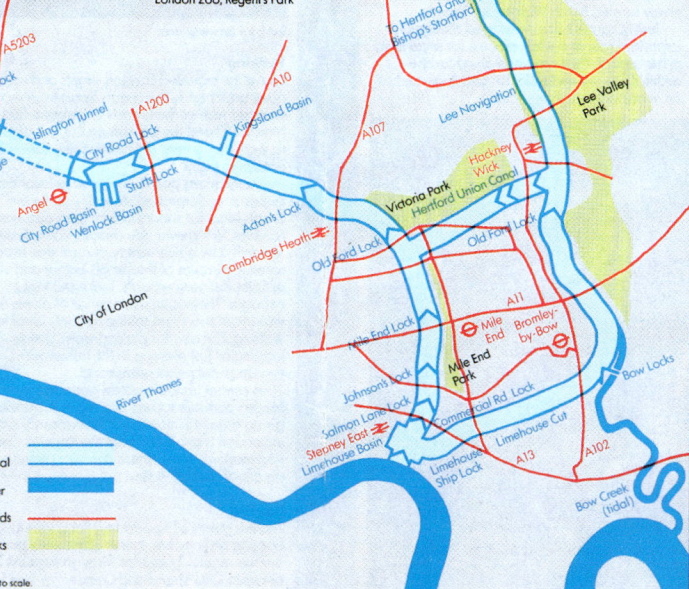

17

18th-19th Century Water Companies

The Pioneering Companies

The industrial developments taking place in Britain from the middle of the 18th century not only increased the demand for water for human consumption, but also expanded the demand for water in manufacturing processes and as a source of power in the form of steam. Steam engines had been developed for the purpose of pumping water. Furthermore the rapid growth of the urban population was soon to provide both pressures and incentives for improving and expanding the supply systems. Private investors hoped to secure a good return from their capital investment.

In 1712 the York Building Water Works became the first to use steam engines for raising water. The Chelsea Water Works Company was formed in 1723 and two years later a Swiss visitor commented that one of the conveniences of London was that everyone had an abundance of water.

During the 18th century there were many contractors repairing old pipes and laying new ones. Their work was not co-ordinated which gave rise to problems in the streets as soon as one contractor completed laying pipes, another would follow, digging the same roads and streets. In a parliamentary debate on poor street surfaces in 1728 an MP observed that the public companies for raising the Thames water were perpetually laying down their pipes or amending them and such a Bill would prove of no purpose. The water companies responded and in 1746 the Chelsea Water Company laid the first permanent iron pipes to replace those of wood and lead.

London Bridge Waterworks

The New River Company did not stop the London Bridge Waterworks from continuing its operation. The first two water wheels of Peter Morice were destroyed in the great fire of 1666 but were replaced with four new ones. However, with the considerable increase in the number of ships arriving in the Pool of London during the 18th century, the London Bridge piers and wheels became a source of hazard to shipping. The eminent civil engineers of the day, Brindley and Smeaton, designed special water wheels to mitigate this situation. But this was short lived and by Act of Parliament, the water wheels were dismantled and the company taken over by the New River Company.

On the south side of the Thames St Saviour's Mill supplied water to parts of Southwark. It was driven by Thames water falling into a pond in St Mary Overies which later became a dock. For about 150 years, the building and the dock, as part of Hibernia group wharves, belonged to the Hays Wharf Company which operated on the east side of London Bridge. In the early 1980s, Hays Wharf and dock were converted into Hays Galleria with fashion shops, cafés, etc. Mr Hays first opened his brew shop on the same site during the 17th century!

The Southwark Water Company was formed in 1760 and Lambeth Water Company in 1785. Other companies followed at the beginning of the 19th century including the Vauxhall Water Company 1805, the West Middlesex Water Company 1806, the East London Water Works Company 1807 and the Kent Water Works Company 1811. With the rapid increase in the number of water companies in London competition became fierce.

Sand Filters

Most of the water supplied by the water companies came from the River Thames, but there was concern regarding the quality of this water. The publication in 1827 of John Wright's article "The Dolphin or Grand Junction, Nuisance" exposed the fact that the Grand Junction Water Company in Chelsea drew its water from the Thames close to the outfall of a major sewer. A Royal committee was established in 1827 to enquire into the sources of London's water supply. The committee recommended the supplies should be taken from other sources than those which were in existence at the time. In 1829 the Chelsea Water Works Company consequently announced the introduction of slow sand filters. This first stage of purification of water was adopted later by other companies. In 1830 another select committee of the commons was formed to survey alternative sources of supply. As their report was being prepared London's first Cholera outbreak occurred in 1831. Unfortunately the cost of providing alternative sources of water was high and the report of the committee was shelved.

The Polo Commissioners Sanitary Report of 1842 was a strong one describing the living conditions of the poor and the inadequacy of drains and water supply. It recommended a constant water supply to every house with improved sanitation and the cost was calculated at 2d per week per house. In 1849 Dr John Snow's pamphlet the Mode of Communication of Cholera was published. It stated that Cholera was caused by poison extracted from a diseased body and passed on through drinking water which had been polluted by sewage.

Chadwick's Recommendations

In 1850 Edwin Chadwick's report on supply to the metropolis contained three main recommendations: that the London water supply should be derived from new sources; that the water companies should be bought out on behalf of the public; and that an executive commission should be appointed to administer the combined water supply and drainage services. A Bill was prepared but was floundered by a select committee which was faced with conflicting advice and issues of the private companies. The Prime Minister Lord John Russell then announced that the Government did not intend to proceed with the Act during that session. A letter subsequently published in the Times on 14th July 1851, demonstrated the uncertainty and irregularity of London water supplies. The Great Exhibition of 1851 in Hyde Park required the authorities to supply pure water and glasses to all visitors demanding it. The magazine 'Punch' commented that whoever can produce a glass of water fit to drink will contribute the best and most useful article in the whole exhibition!

In 1852 the Metropolitan Water Act was passed which dealt with some of the problems and abuses of the water companies. Those companies supplying Thames water were instructed to move their intakes above Teddington Lock into the non-tidal part of the River Thames. The companies were allowed three years to implement these changes and move their intakes upstream. The Chelsea Water Works Company was given four years and five years to comply with the other measures. Maximum penalties were specified at £250. In 1853 the Lambeth Water Works Company became the first to move its intakes above the tideway as required by law. Its death rate fell to 37 persons per 10,000 compared with that of the Southwark and Vauxhall Water Company of 130 deaths per 10,000.

Outbreaks of Cholera

After the devastating ravages of cholera during the 1830s and 1840s, the Government intervened in 1856 and established the Metropolitan Board of Works. This Body had control over 117 sq.miles of the capital and Sir Joseph Bazalgette was its Chief Engineer until its replacement by the London County Council in 1889.

Royal Chelsea Flower Show and Hospital showing the Ranalagh Common surface water sewer discharging into the Thames, c1990s.

Punch cartoon by John Bull illustrating the water companies cutting off supply to the ordinary people. There is an absolutely obscene hieroglyphics all round the edges of the picture.

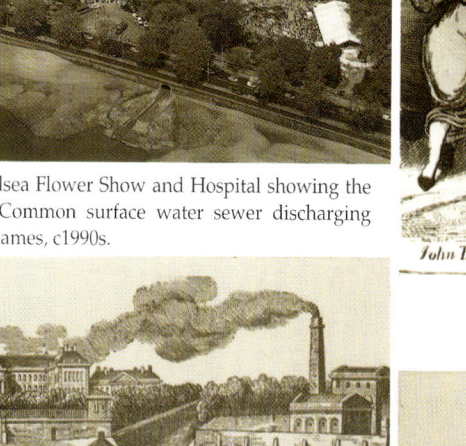

Royal Chelsea Hospital and Ranalagh Common open sewer, c1725. The pumping station (right) was pumping untreated water directly into the mains.

Microcosm dedicated to the London Water Companies. This illustration is from Punch and dates from around 1845.

Top: A water cob with realistic sized water carriers. The hat would suggest middle 17th century.
Right: Cartoon illustrating unreasonable restrictions imposed by East London Water Company in supplying the local people.

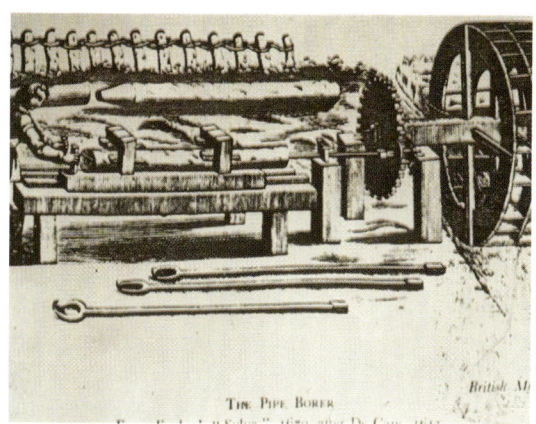

During the 16th & 17th centuries elm trees were hollowed out on the site of today's Hayes Galleria to provide London with its first water mains. The above engraving is called The Pipe Borer taken from Evelyn's "Sylva" of 1670.

Engraving of a water cob in the Middle Ages, using a heavy carrier on both his shoulders.

Southwark and Vauxhall Fire Brigade c1905, showing a horse-drawn pump engine.

Meagre Daily Supply

The problems of constant and adequate water supply had to be resolved. In 1862 the Medical Officer of Health for Whitechapel reported that in 48 out of 133 areas the water supply was only by stand taps which supplied water daily (except on Sundays) for a period varying from 15-30 minutes. In the meantime Sir Joseph Bazalgette's sewage scheme was completed removing one danger of pollution.

Metropolitan Water Board

In 1869 a Royal Commission recommended the introduction of a constant and purified water supply under public control. It was noted by Sir John Simon the colossal power of life and death of the water companies. In 1871 the Metropolis Water Act attempted to remedy these problems. It required the companies to supply water everyday, but mainly empowered the Metropolitan Board of Works to request a constant supply of water to any area and under certain conditions empowered the Board of Trade.

Although the water companies continued to receive adverse publicity throughout the 1890s it wasn't until 1897 that yet another Commission was appointed to review the problems. As a result of this Commission's recommendations, the Metropolitan Water Act was passed in 1902 and by then the Metropolitan Water Board was established over the eight private water companies at a cost of about £40 million. The Board became operational in 1904 and its headquarters were established in New River Company offices. In April 1974 the administration of the Metropolitan Water Board was transferred to the new Thames Water Authority. In 1992 the Authority was privatised and it is now functioning under a number of companies.

Water Supply to East London

The water supplied to East London including Whitechapel and Stepney, was by Shadwell Water Works, set up by Thomas Neale in 1669. He leased land at Shadwell and pumped river water using horse driven pumps. Later the company was incorporated and Neale became the first governor. In 1747 steps were taken to supply water to Stratford, West Ham and Bow, from a pumping station on a branch of the River Lee, the pumps being worked by a waterwheel. Both Shadwell and West Ham Water Works were purchased in 1807 by the London Dock Company and this is perhaps the first link between water supply and London Docklands. The situation was short-lived because the East London Water Works Company purchased them from the Dock Company later that year.

At about the same time as the West India Docks expanded trading, the East London Water Works began operations from Old Ford where they constructed a sedimentation reservoir. Water was drawn from the River Lee and allowed to settle in the reservoir. The principal aim of the water companies at this time was to supply water which looked reasonable and so sedimentation was the only treatment. The company was supplying water to 10,739 homes in 1809 and an indication on the growth of London Dockland following the construction of the docks is shown by the fact that in 1819 the number of houses supplied had risen to nearly 30,000. In 1829 the company purchased the Hackney Water Works in Leabridge Mills. They also obtained an Act of Parliament empowering them to move their intake works upriver to Lee Bridge where non-tidal water was available.

Although public water supplies were being treated by filtration as early as 1829 it wasn't until 1853 that the East London Company introduced slow sand filtration to Leebridge Works. This was a result of the 1852 Metropolis Water Act which prevented abstraction of water from the Thames below Teddington. It should be noted that even at this time the East London Water Company's boundary did not include the Isle of Dogs. During 1853 the company was authorised to make new cuts in connection with the River Lee for improving the water quality and to construct impounding reservoirs at Walthamstow as well as the filter beds at Lee Bridge.

The last outbreak of cholera in London was in 1866 and was wholly confined to the area supplied from the Old Ford Works of the East London Water Works. In 1867 the company was given powers to construct a surface reservoir at Finsbury Park and to build abstraction works on the Thames at Sunbury and Hanworth. The water was pumped to Finsbury Park reservoir via a 0.9m diameter main which is still in use today.

Despite all of these improvements the problems of the East London Water Works Company were not over. During the 1890s there were several years of drought in East London when water was rationed. This was arranged by turning the water on for six hours per day only and led to a great outcry of complaints against the company. The era of the water companies was rapidly drawing to a close and this was achieved by the formation of the Metropolitan Water Board in 1904.

The East London Water Works Company had prepared Acts of Parliament in 1900 to allow them to expand the number of reservoirs in the Lee Valley. The Board continued with the works and the first reservoir, named after King George, was opened in 1913. However, the second reservoir, William Gurling, was not opened until 1951.

Water Supply Illustrations

Illustrations on the water companies are shown on the following page. Further details are given below.

The Waterworks of London Bridge for the supply of the City of London with Thames water was the first mechanical scheme for the Capital. The diagram on page 9 shows one of the waterwheels with lifting apparatus, taken from the Universal, May 1794. London Bridge Water Company's pumps were first installed around 1570-80. They were mounted in the first and second arches of the old stone London Bridge. The weir effect either side of the bridge produced a difference in water levels of about 2 metres to drive the wheels. The drawing is a representation of the wheel exploded because everything is all concealed. The four levers on the left were the actual pumps although very small in comparison with the apparatus and would not have pumped a lot of water. The Company building and wooden water tower were on the embankment.

London Bridge retained its medieval appearance until the middle of the 18th century. It was designed in 1176 by a clergyman, Peter de Colechurch and constructed in masonry to replace a succession of timber bridges first built by the Romans in 72 AD. It was replaced in 1209 and remained as the capital's main river crossing for 600 years. Despite the ravages of the fire of 1666, its houses and shops remained its most distinguishing features for most of its life. Toll was charged for crossing the bridge. By the early 19th century the medieval look of the bridge had disappeared. The water wheels of the London Bridge Waterworks, first installed in 1582, were replaced by George Sorocold in 1702 and later added to. They relied on the bridge piers to narrow the flow and create a water head of about 2 metres to drive them.

The York Buildings Company was set up in 1675 to supply water to St James, Whitehall and Piccadilly. Horse power was initially used to pump water from the River Thames south of Fleet Street to the top of the Wooden Water Tower where it was distributed to customers. From 1713, the early Thomas Savory's steam engines were used. The Stone Watergate can be seen today near the Strand.

The Chelsea Waterworks Company was established to supply Westminster district around 1726. Initially the rising tide used to drive the pumps but this system was later replaced by two steam engines built by John Wise. The company had service reservoirs in St James and Hyde Park. In the 1820s their engineer, James Simpson, developed sand filtration technique for the purification of water. Present day Victoria Station was built where these works used to be.

Watercolour of the New Riverhead Windmill tower in Anwell Street, looking up towards Pentonville Road. Note the New River to the west of the picture, c1760.

A general view of the New Riverhead Waterworks looking south towards the City of London taken from an eminence near Islington, c1768.

View of London Bridge early 19th century. The London Bridge Waterworks Company was started in 1582 by a Dutch engineer, Peter Morice, to supply the City of London.

Water pipe laying in Drury Lane c1845. The cast iron pipes, 0.6metre diameter, would have been all lead jointed and laid approximately 1m deep. Note the first gas main laid in 1810.

Watergate and York Buildings by Malton, c1795. The York Buildings Company was set up in 1675 to supply water to St James, Whitehall and Piccadilly.

Water pipe laying in Piccadilly c1836. Old Bond Street is to the left of the picture. Note the first gas lamps for street lighting.

Chelsea Waterworks, an engraving by Brydeck, c1752. Around 1726 The Chelsea Waterworks Company was established to supply the Westminster district.

Water pipe laying in Tottenham Court Road, c1834. The jointing lead was melted on the left of the picture.

21st Century London Water

London Ring Main

To help secure a supply of constant and safe water for London during the 21st century required a radical solution. London's water is fed from a network of mains, buried just under the ground's surface. Using these mains water is then pumped across various parts of the city from the major treatment works situated on the Thames and the Lee Valley on the outskirts of the capital. These works have to be situated away from the tidal Thames partly because the water below Teddington Lock in Middlesex mixes with sea water as a result of the tidal nature of the river. This system is energy intensive as it requires not only continuous pumping of water over long distances but often repumping once or even twice to reach the outmost extremities of the pipe network system.

In the new Water Ring Main huge tunnels of an average diameter of 2½ metres have been excavated under London. They have been driven at an average depth of 40 metres below the surface of the city. They move underneath most other tunnels such as telecommunications and underground railways. The Ring Main supplies about 50 per cent of London's present day demand for water. Water is raised from twelve shafts located around London from Ashford Common in the west, to Copper Mills in the east (see diagram) where it is then pumped to the customer in the normal fashion. The vast majority of equipment is installed underground and unseen which helps to protect the environment of the City. For instance it is hard to believe that in the middle of Park Lane there is one of the major pumping stations for London as there is no building above ground at this site. In the shafts below ground electric pumps lift the water that has already been treated at the works, from the tunnel and pumped directly into the supply ready for the public to use and drink. All the shafts are unmanned and controlled using up to date computer equipment from a central control centre.

Access to the shaft is only for essential maintenance. The shaft tops are tightly secured and the shafts themselves are dry. They contain the valves and rising mains which are sealed under pressure. And within the tunnel itself fibre optic cables carry the signals which remotely monitor flows, pressures and controls the pumping system. The concept of the tunnel is simple. Water enters the tunnel from the treatment works and these have the capacity to ensure the tunnel is always full. So if the supply in one works runs low, then treated water from another moves to take its place. This is accomplished by using one of natures most powerful forces - gravity which is free. Finished in 1996 the London Water Ring Main is able to move 1300 mega litres (285 million gallons) per day. The tunnel and further extensions could increase its length to 140 km in the 21st century equivalent nearly to two Channel Tunnels in length.

The New River 2000

During the new millennium the nearly 400 year old New River, is a key to the recharge scheme for London. The huge underground aquifer which was once a primary source of London's water, was over pumped during the 20th century almost to extinction. In the early 1990s a scheme to recharge artificially the huge chalk aquifer in North London by pumping water into it from the New River has proved successful and is used as part of Thames Water's advertising campaign on television. The spiralling demand for water and the knowledge that in times of drought, the modern supply schemes cannot satisfy the capital's thirst.

The plan is to use the aquifer as an underground reservoir by recharging in times of plenty and during drought to use it to boost the supply. In total 23 bore holes are in place along the River from Turkey Brook to Green Lanes. Of these, eleven at the upper end are able to recharge and abstract with the remainder, initially, only used for abstraction. When the main system has a surplus, water from the distribution system will discharge into the aquifer via the upper boreholes. During drought, the aquifer will return the compliment.

New River Head Today

The New River Head building at Islington was originally commissioned in 1913 as the headquarters of the Metropolitan Water Board and completed seven years later. It is a monument to municipal grandeur with a glazed barrel vaulted revenue hall and the old Oak Room panelled in wood carved in the 17th century by Grinling Gibbons. It passed into the possession of Thames Water in 1973 and became surplus to their requirements when the Authority was privatised into a company in the 1990s. In 1996, developers converted the building into 130 apartments which were sold between £150,000 - 600,000.

The New River Head has many original features which have been retained. Hardwood screens surround the revenue hall, where water rates used to be paid, curved plaster arches, frame windows in the upper floors, cornices and panelling feature throughout the building. English Heritage has specified the items to be preserved, including a group of wooden telephone boxes and a two-storey boardroom. The barrel vaulted hall is retained as a communal space.

The vaulted barrel revenue hall of New River Head is being renovated as a communal space, c1996.

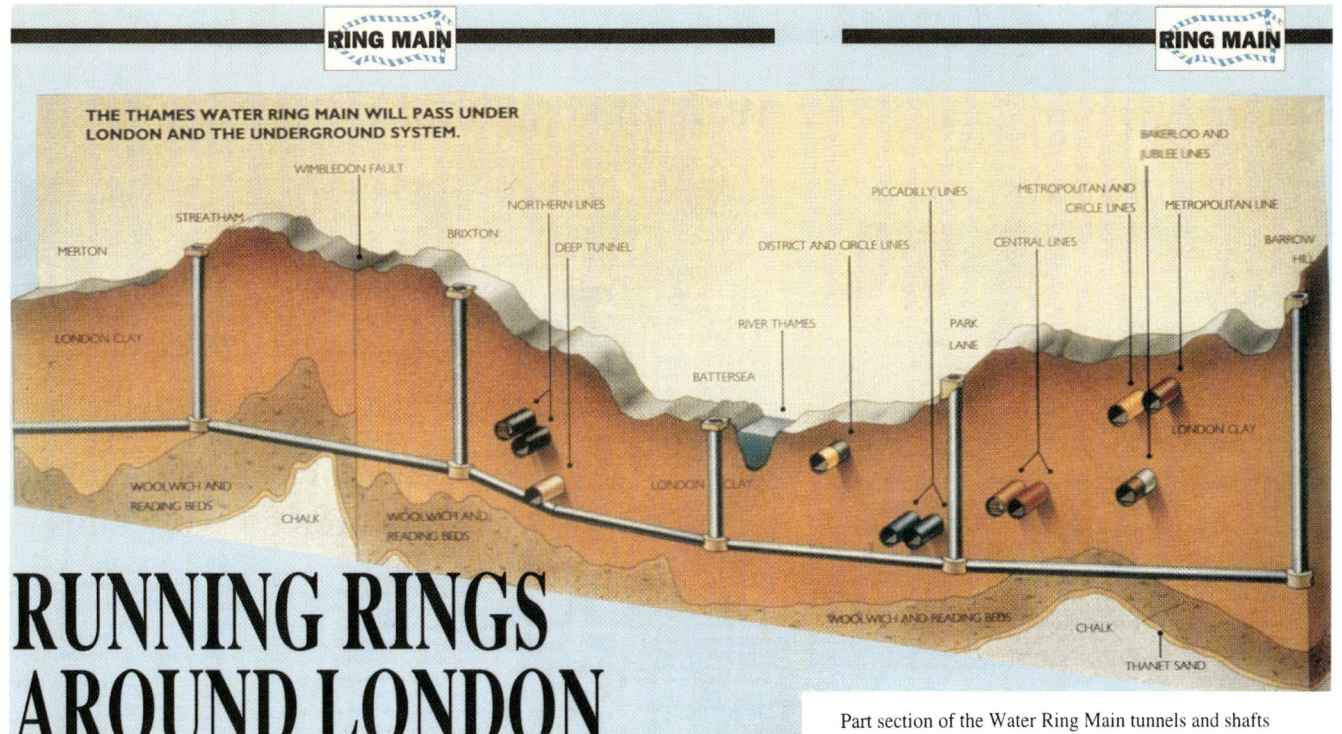

RUNNING RINGS AROUND LONDON

Part section of the Water Ring Main tunnels and shafts under London, circa 2000.

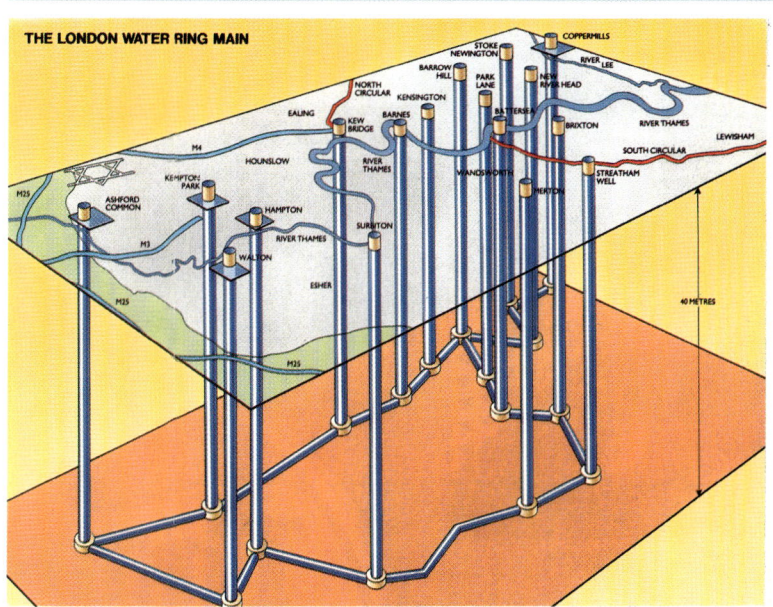

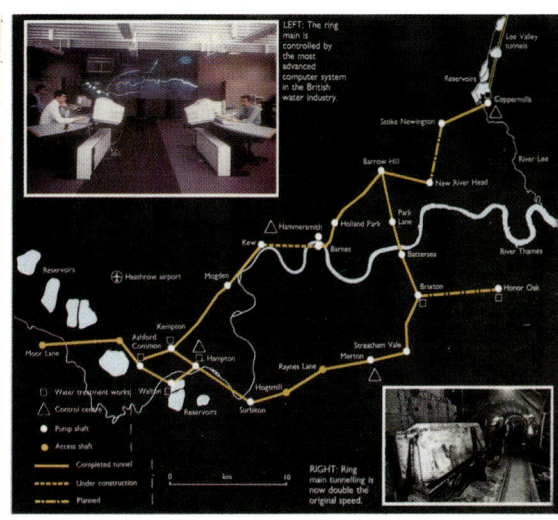

Top: Plan of the Ring Main, River Thames and water storage reservoirs in London.
Left: A three-dimensional view of the Ring Main underground.

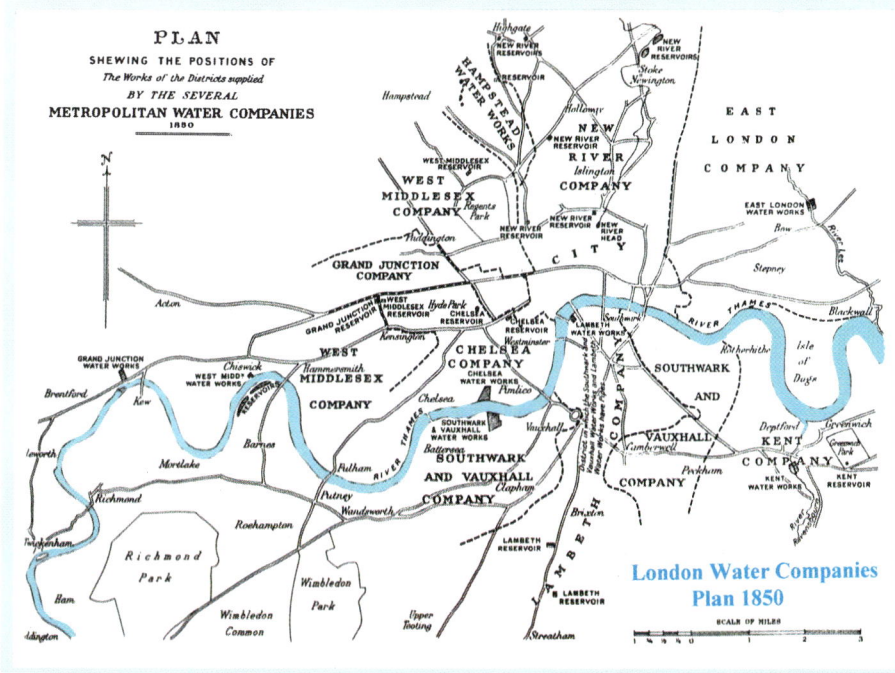

London Water Companies Plan 1850

Top: A bird's eye view of the top of the covered Park Lane shaft, near Hyde Park Corner, circa 2000.
Left: Plan of Victorian London showing the private Water Companies and the boundaries of the districts they served in the middle of the 19th century.

London Main Drainage

Under ground London
London has undergone dramatic changes over the past three decades. Building development has taken place above ground but it has also developed a complex system of services below ground. Beneath the City's streets gas, electricity, water and sewerage services mesh with telecommunication cables, transport tunnels and piled foundations of post-war office blocks. Much of the existing infrastructure is being upgraded and the work will continue well into the 21st century as much still remains to be done. The following is a brief account of its drainage story.

Medieval Times
The gullies running down the middle of the cobbled streets were used as sewers where rubbish and other sewage were washed away by heavy rainfalls into ditches, streams and finally the River Thames. Cesspits were dug around the houses and their contents and those of earth closets would be carted away to open public laystools and this could be carted again to the market gardens and fields as manure. Although the water closet had been invented by Sir John Harrington as far back as 1596, it could not be generally used until adequate drains and public water supplies became available.

The tributaries and the River Thames were open sewers in Medieval times and a petition to Parliament was made in 1290 by the White Friars who declared that the putrid air around the Fleet overcame the aroma in the incenses and it caused the deaths of several brethren. Until it was covered over and used as a sewer the Fleet River was always a smelly ditch. In 1427 came the first of a series of Acts appointing local commissions of sewers but they were mainly concerned with the drainage of surface water and the prevention of flooding. In 1531 an Act was passed that remained statutory until 1848.

Metropolitan Improvements
The City of London had no properly planned system of drainage until the Metropolitan Commission of Works was established in the middle of the 1840s. With the enormous expansion of the metropolis during the Victorian era and the growth of the industrial revolution, and finally the re-occurring outbreaks of cholera required radical reforms to be carried out. The Metropolitan improvements of around 1830 during the famous architect Nash's time, laid some sewers below the streets and Nash himself built a fine one which still serves below Regent Street. Yet these did little to improve the general situation. Such sewers were mostly designed to carry away rainwater and the discharge of excrement or garbage into them was indeed an illegal offence often ignored until well after the mid 19th century. The general concern at the time was expressed by Pepys in his diary where he records how his wife stooped into the street "to do her business".

In the reconstruction of the City after the Great Fire of 1666 the rebuilding Act of 1667 gave wide powers of altering, enlarging and cleaning the City sewers but this lapsed when the rebuilding was complete. The Metropolis General Paving Act of 1817 provided for the cleansing of drains and cesspits but ten years later the Westminster Commission of Sewers stated that the care of road drainage was the only matter of which the public was concerned the rest being the concern of property owners. The Metropolitan Building Act of 1844 made it mandatory for the collection of all the drains to sewers in new buildings. The census of 1841 had revealed that over 270,000 houses stood in the Metropolis, most of which had a cesspit below.

In an attempt to rectify this hideous situation the Metropolitan Commission of Sewers was formed in 1847 to amalgamate the eight separate local bodies normally responsible for London's drainage. It decreed that all cesspits must be abolished and so 200,000 cesspits suddenly went out of use. The effect was disastrous for now all the main sewers and underground streams discharged their new contents into the Thames which was compelled to accept the sewage of three million human beings in the City. The river became a huge open sewer. In 1800 salmon had still been swimming up to London and beyond but by the mid 19th century by which time the population had doubled, no fish of any kind could survive in the river and even the swans had deserted it!

In the hot dry summer of 1858 the climax arrived in what came to be called the Great Stink, when the windows of the Houses of Parliament had to be draped with curtains soaked in chloride of lime to mitigate the disgusting smell! Tons of chalk lime, chloride of lime and carbolic acid were tipped into the river with little effect and Disraeli described the Thames as a "stagnant pool reeking with unbearable horror". The worst effect of this improper drainage was not the stink but the appalling cholera epidemics that spread in London for three decades most of which occurred in the poorer districts. It was then thought that the Asiatic cholera was caught by the inhaling of noxious vapours but by then Dr John Snow proved his theory that cholera was due to the contamination of drinking water. Not until 1883 however was the microbe of cholera isolated by Robert Koch but by the year of the great stink radical improvements were imminent.

Mid 19th Century London
Despite these problems London in the 1850s must have been an exciting place in which to live. The population was rapidly expanding as their commercial, industrial and government roles increased. The Victorian city grew wealthy as the centre of a large British Empire stretching in five continents. It was nevertheless becoming insanitary with water and sewerage systems dating from medieval times. It was apparent that drastic sanitary improvements were required. The Metropolis had suffered severely from a number of cholera outbreaks resulting in eighteen thousand deaths in 1849 and twenty thousand in 1854.

It was in 1853 that the civil engineer, Mr Joseph Bazelgette, came prominently to the scene. He had since 1845 held office under successive commissions. On the death of his Chief Engineer, Mr Frank Forster, he was appointed as his successor and instructed to prepare a drainage scheme for the Capital. Progress of his scheme submitted in 1854 was frustrated by further changes in the Commission resulting in the formation of the Metropolitan Board of Works. Its first act was to appoint Mr Bazelgette as their Chief Engineer. It might have been assumed that under these circumstances the construction work would at least start. This was far from the case and many questions were raised regarding the details of the scheme. Meanwhile the state of the river was becoming worse and Parliament was in despair at seeing nothing being done.

Act of Parliament for Scheme
In August of 1858 the Prime Minister, Mr Disraeli, succeeded in passing an Act of Parliament which relieved the Board from Government sanction on the grounds that as the Metropolis paid for the works, they had the right to construct them in any way they pleased. The First Commissioner expressed it afterwards: "He who pays the fiddler has the right to call the tune!". This paved the way for Bazelgette to complete the design for the whole of London and contracts for the construction work were arranged.

Sir Joseph Bazalgette, the great civil engineer, built the London Drainage system in the 1860s and put a stop to widespread outbreaks of cholera in Victorian London. Among other works were the river embankments and Thames bridges still in use today.

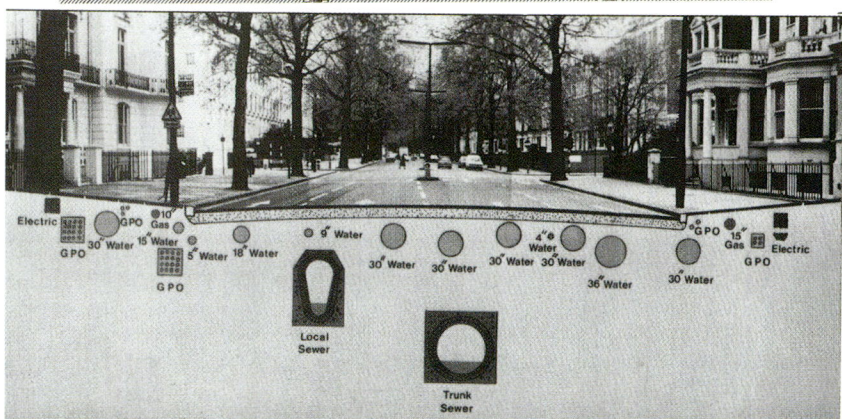

Fleet river in the City used as a sewer c1845.

Top Left: 19th century engraving of water mains, sewers and railway tunnels.
Middle: Present day underground services in Holland Park Road, West London.
Bottom: Typical Victorian sewer built by Bazalgette.

Sir Joseph Bazalgette Memorial (1812-1891) - Chief Engineer of the London Drainage System and the Victoria Embankment.

Crossness Beam Engine House being converted into Heritage Museum, 1990s.

Interior of the Beam Engine House under renovation, c1990s.

The Rat Catcher of the Sewers - from "London Labour and the London Poor" written by Mayhew, 1849-51.

Abbey Mills Pumping Station on the northern outfall is currently being renovated for other uses.

The interior of Abbey Mills Pumping Station around 1868, from the Illustrated London News.

Southern Outfall works where the southern sewer discharged at Crossness in Kent, c1865.

The interior of the Crossness pumping station engine house on its opening by the Prince of Wales on 4th April 1865.

Repairs of the Fleet sewer circa,1890. The sewer was dried by building two dams and water diverted through a wooden channel above the level of the dams.

The Victoria Embankment under construction showing the sinking of sheet piles on one side and then filling in. The Acts to build the Embankment were passed in 1860.

Typical present day sewer under London dating from the 1860s which has recently been renovated.

Inspection of a Victorian cathedral-like sewer junction with storm overflow weir on the right.

Wick Lane Depot under construction. The two arches are the sewer although they look like a roof. The reason for this is that the outfall comes out of the ground and is all embankment.

The scheme for the removal of sewage was staggering in its scope in preventing any sewage passing into the Thames in the London area. New lines of sewers were constructed at right angles to the existing and a little below their levels so as to intercept their contents, then conveying them to outfalls downstream. At these outfalls, the sewage was delivered into reservoirs placed at such a level which enabled them to discharge around the time of high water. By this arrangement the sewage was not only diluted by the large volume of water but was carried out on the ebb tide to a point some 42 kilometres below London Bridge.

Opening of Drainage System

On 4th April 1865, the Prince of Wales opened the Southern Outfalls Works at Crossness on the lonely Belvedere Marshes. The Victorian engineers took great pride in their public works, providing visible "castle like" buildings as evidence of service to the community and Crossness was one of these. The main steam engine house was built in Italianate style with a chimney stack akin to a companile to serve the boilers. Inside the building James Watt and Company installed four engines named "Victoria", "Prince Consort", "Alexandra" and "Albert Edward". Their function was to pump the sewage from the main sewer into the reservoir for eventual release. The steam engines were condensing double-acting rotary beam engine types and the pumps reciprocating plunger or ram pumps with the sewage being discharged through a series of valves. It was claimed that the contractors for the Crossness engines guaranteed the engines would raise 80 million pounds (0.45 kg) one foot (0.3m) with one hundredweight of Welsh coal! The finish of the engine house interior was quite remarkable. The structural cast iron work for the columns and screens of the central octagonal shaft connecting the three floors of the building shows a mastery of structural steel nearly 150 years ago. The total cost of this huge scheme was also remarkable at £4.6 million.

The scheme consisted of a network of sewers at three levels on both sides of the river running down to the lowest level of outfall sewers that ran eastward to reservoirs some twenty six miles below London bridge on the north at Beckton just west of Barking Creek and on the south at Crossness to the north of Erith Marshes. Through the outfalls both sewage and surface water are discharged except during excessive rainfall when the surface water can be run straight into the river. The high and middle level sewers discharged by gravity while the low level ones are aided by pumps built at a number of points.

The other impressive pumping station, Abbey Mills in Newham, was designed in Gothic style by Bazalgette and construction was completed in 1868. the beautiful Victorian building is still in use today and has recently been renovated.

When finished Bazalgette's scheme consisted on 1300 miles of sewers mostly built of stock bricks, 82 miles of which were main intercepting sewers. At Deptford, Stratford, Pimlico and Crossness rose the pumping stations of romantic design and masterly engineering which are now a monument to the great Victorian drainage age. Another of Bazalgette's achievements was the building of a number of London bridges and embankments including the Albert, Victoria and Chelsea embankments and he incorporated a length of the northern low level sewer in the Chelsea and Victoria embankments.

Thames Water Authority

London's drainage has been increased and improved greatly as London has grown but Bazalgette's work remains at its core. In 1974 the Thames Water Authority replaced the Greater London Council and local authorities in controlling London's 800 miles of main drainage when it also assumed control of the entire water cycle of the 500 square miles of the Thames basin and catchment area. London's complex sewage system now treats over 2700 million litres a day at fourteen treatment works. The purified effluent flows into the Thames and its tributaries and the sludge is rendered inoffensive. The river grows cleaner year by year and in 1974 the first salmon in the river for 150 years was found trapped in the filters of the West Thurrock Power Station.

Bazalgette Achievements

Bazalgette, did not get the credit he deserved although he was eventually knighted and had his modest monument on the embankment beside Hungerford railway bridge. The son of a Royal Naval Commander of French extraction, he joined the staff of the Metropolitan Commission of Sewers on its formation in 1845 and was appointed Chief Engineer to the new Board of Works. He at once began preparing his grand drainage scheme for London. Although the Prince of Wales declared part of it open in 1865 the whole scheme was not completed until 1875. Its effect on the health of Londoners was dramatic. The cholera epidemics were over for good.

Present day Londoners owe a great deal to the genius of Sir Joseph Bazalgette, whose contribution to the capital is unique. In reply to comments on his paper, "The Main Drainage of London", presented to the Institution of Civil Engineers in March 1865, he said he now found there were a great number of persons who claimed to be "pioneers" of his design. It was a source of great regret to him that he had not known of this before as it would have saved him many anxious days and sleepless nights.

Drainage Illustrations

Further information on these illustrations are given below.

Thames Victoria Embankment - One of the pictures on the previous page is the Victoria Embankment under construction looking east along the river with St Paul's Cathedral and the City in the background. The bridge in the distance is Blackfriars. The construction picture shows the sinking of sheet piles on one side and then filling in. They are also building the District Line of the underground railway under the embankment, circa 1861. The Acts to build the Embankment were passed in 1860.

Sir Joseph Bazalgette Memorial - Chief Engineer of the London Drainage System and the Victoria Embankment. (Born 1812 and died 1891), was a penny pinching grudgingly provided tribute by the London County Council. The Metropolitan Board of Works finished up with an air of corruption and the LCC came in with the intention of improving things. Apart from his contribution to London drainage which is generally regarded as unique, he changed a lot of the practices for which he was responsible. The memorial was erected in 1899.

Chelsea Hospital and the Ranalagh Common sewer discharge circa 1725, is in use as a tunnel today (see page 20). The first picture reveals something which has always perplexed people as to why the Chelsea Hospital is askew to the river; you would think it faced directly to the river, but now it is at an angle. The old engraving shows that it was square to a big bend in the river. That bay was straightened out by the construction of the Chelsea Embankment thus leaving the hospital at an odd angle. At this time sewers were only used for the drainage of surface water and it was illegal to put sewage into the rivers. It is interesting to note that the intake and the sewer discharge were situated at the same place west of the pumping station. Water was pumped directly into the mains completely untreated!

One of the Embankment 'mooring' hooks.

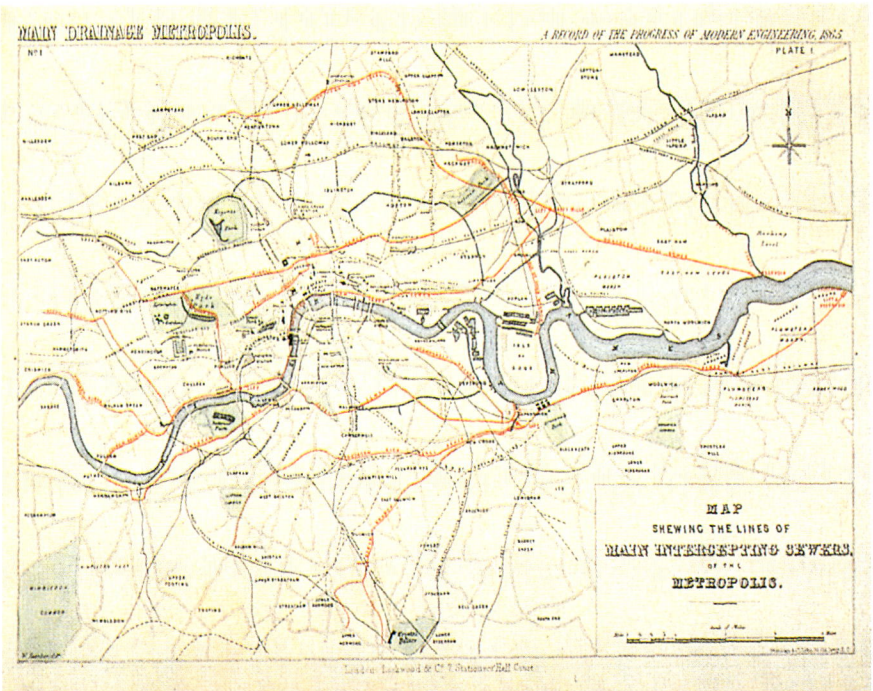
Plan of London showing the six main interceptor sewers designed by Bazalgette.

The pumping station at Pimlico, still in use, was designed to look like a Venetian palace.

View of Beckton Sewage Works looking towards the Thames and Barking Creek.

Abbey Mills, the impressive gothic pumping station.

Restoration work inside Crossness pumping station.

Detail of a mid 19th century map showing street sewers.

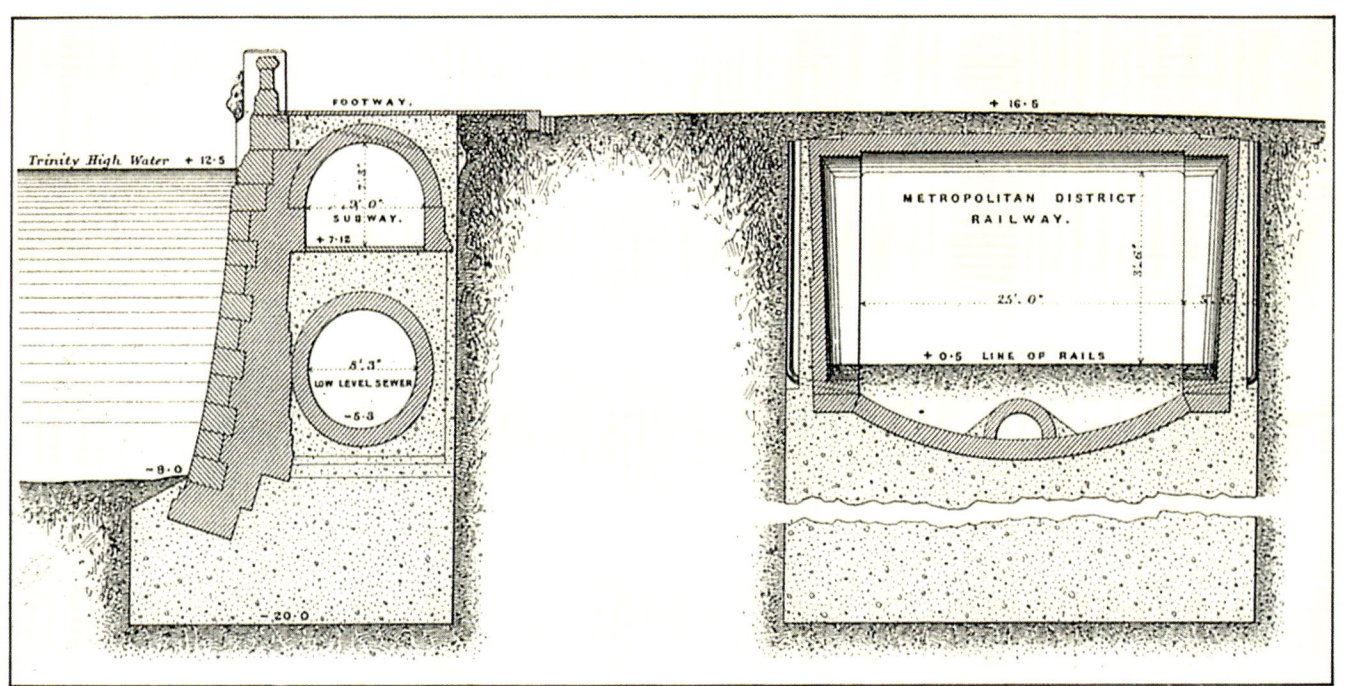

Cross section through the embankment at Charing cross showing the new river wall, sewers, and Sir John Fowler's Circle and District Railway.

Aerial view of Crossness Sewage Works.

The Victoria Embankment opened up the banks of the Thames as a public amenity.

Discover the River Thames

Hampton Court to Richmond

Starting from West London, Hampton Court is a splendid landmark along the River Thames. The old Tudor Palace has a significant spot on the river for it indicates the place where suburban London may be said to begin. The multitude of towers and mullions diversify the façade while the graceful clocktower and high-pitched roof of the Great Hall break the skyline beautifully.

Between Thames Ditton and Surbiton the river has a broad reach with the picturesque Thames Path on the north side. Kingston is a very interesting old town with considerable changes over the past two decades. It was an important place in the scene of coronation of Saxon Kings during the first Millennium. At Kingston the Thames Path changes to the south side and the river takes a bold sweep to Teddington.

At Teddington the weir is the prime delight of any angler on the river. Adjoining this is the Teddington Lock which is the last of its kind in the Thames. The lock and weir mark the spot, between 96-112 kilometres from the sea, at which the Thames seems to be tidal. At Teddington Lock the river winds itself past Eel Island to Twickenham. Nearby is Strawberry Hill, a masterpiece of architecture by Walpole and definitely worth a visit.

Ham, with its famous walks, lies low and little of it can be seen from a boat, but from the Thames Path there is a pleasant glimpse of Ham House with the splendour of the surrounding elms. Past Ham, the river winds through Petersham with the wooded Richmond Park in the background.

Richmond to Battersea

Richmond was once a favourite place of many Kings of England. After the town of Isleworth which lies on a concave bend of the river has been passed, the Thames is bordered on either side by a park. On the north side are the Royal Botanical Gardens at Kew and on the south side the Old Deer Park. Syon House which is separated from the river by a meadow, has beautiful gardens and a nursery.

At Brentford the River Brent makes its way to the Thames from the uplands of Hendon. It is believed this is the spot where the Romans crossed the River Thames in 43 AD.

From Kew Bridge to Putney Bridge the Thames follows a course of double loops forming an almost regular S the axis of each fold lying nearly north and south. On the north side lies Chiswick with its beautiful cottages and houses along the riverside. A short walk from the river is Chiswick House which was the favourite residence of the Duke of Devonshire.

Between Putney and Mortlake, a distance of about 6½ kilometres, is the course of the annual University Boat Race, held since 1870. The race normally takes place on the Saturday before Palm Sunday each year. Putney and Fulham are united by a bridge across the river and on the bank is the attractive Fulham Palace.

Battersea to Lambeth

It is at Battersea and Chelsea that the Thames first acquires the character of a metropolitan capital city. The river here makes a somewhat abrupt curve and gives an outline to the whole locality. On the north bank the Chelsea Embankment has a clear run to Central London. It was towards the close of the 17th century that Chelsea first became socially famous as a pleasant place to live outside Central London

World Famous Landmarks

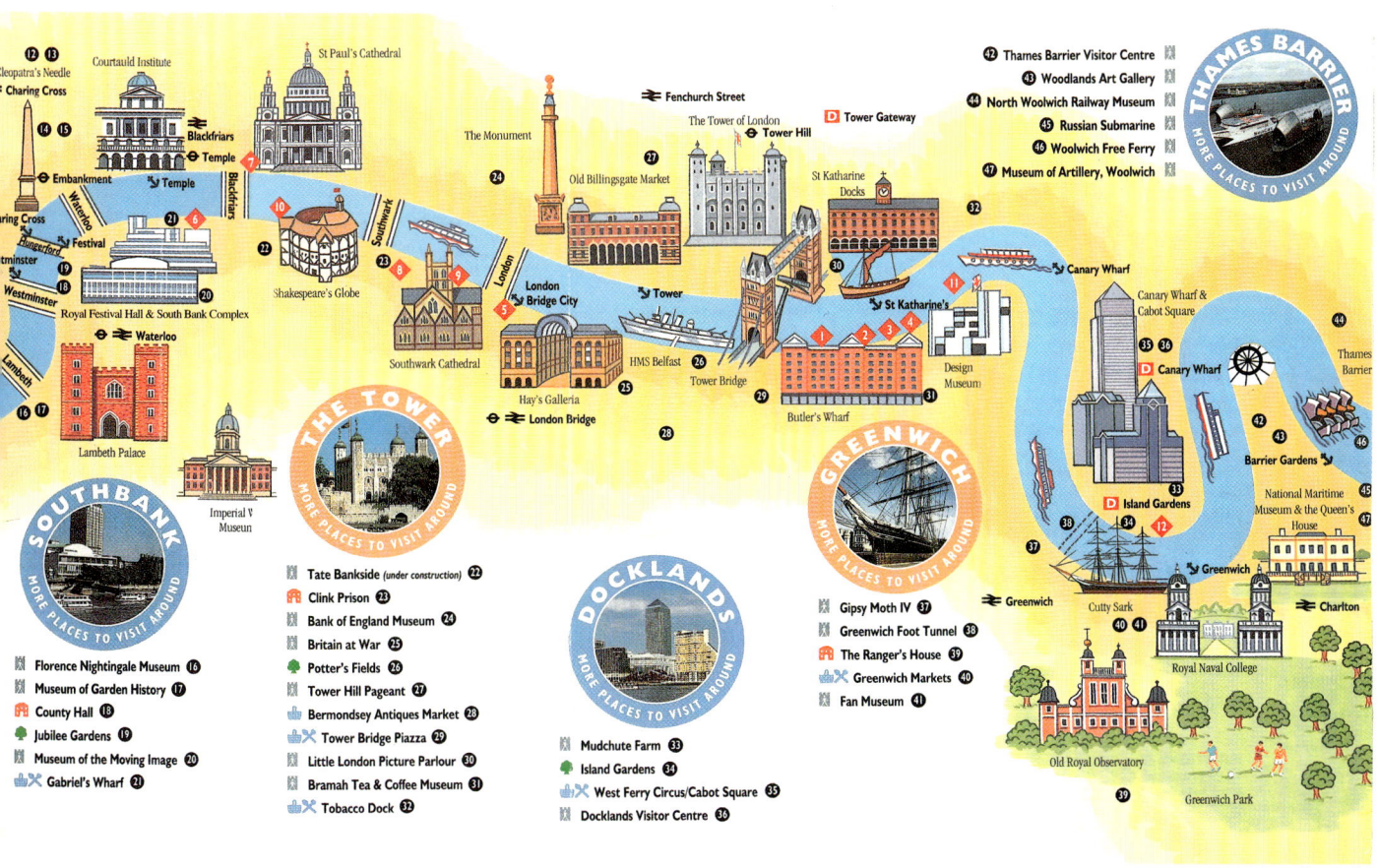

and some of the existing houses belong to that period. Cheyne Walk is the most noticeable road in the area. Further along you will see the Royal Hospital and its frontage is used annually for the Chelsea Flower Show which is known all over the world.

With the hospital grounds on one side and Battersea Park on the other, we come to the Chelsea suspension bridge. The most prominent landmark at Battersea Reach is Battersea Power Station which is awaiting conversion.

Past Vauxhall Bridge we come to Lambeth Bridge. To the walker along the Albert Embankment on the south bank of the river the appearance of Lambeth Palace is an interesting site. It contains the Archbishop's residence and adjacent church. The river walk continues along the Albert Embankment until you reach St Thomas' Hospital near Westminster Bridge. Looking across the Thames from the Albert Embankment we get the best view of the Houses of Parliament and Big Ben. We are facing the ancient City of Westminster with its towers and steeples. At the southern end of Westminster Bridge stand the former County Hall and the new gigantic Millennium Wheel.

London West End
On leaving Westminster Bridge along the north bank the Victoria Embankment begins with a magnificent work of engineering. The massive river wall has bronze heads of lions staring out of each pier, the extended line of parapet, the artistic lamp posts and the Cleopatra Needle with sphinxes around its base. There are the elevated terraces of some buildings and the stately river frontage of Somerset House.

The new railway station at Charing Cross is an impressive building overlooking the river. It was completed in the late 1980s. At the bottom of Buckingham Street, The Watergate, built by Inigo Jones for George Villiers, Duke of Buckingham, still stands along the river in isolation. Waterloo Bridge can be seen next followed by the Temple home of the Legal profession.

The City of London
We are near Blackfriars Bridge which is also the beginning of the City of London water frontage. It is obviously a city of commerce and finance with its incalculable riches.

Although St Paul's Cathedral is some little way from the Thames, its splendid cupola is so prominent an object it is impossible to resist the temptation of a short detour to visit the great Cathedral. During 2000, the New Millennium Bridge was constructed connecting St Paul's to Bankside.

There is no Thames Path further downstream until you cross Southwark Bridge to the south bank. A pleasant part of Southwark is occupied by the new Shakespeare Globe Theatre fronting the Thames.

Returning towards the river we find, not far from the water edge, the old Church of St Saviour., currently called Southwark Cathedral. We are now at the southern end of modern London Bridge, one of the best-known bridges since Roman times. The bridge also marks the western boundary of London Docklands.

Docklands
For detailed descriptions of the river side in the regenerated London Docklands please refer to the author's book on "London & Docklands Walks" (page 64).

New Docklands Waterside

London New Riverside

Extensive developments along London's riverside have been underway since the early 1970s following the closure of the upper docks. These comprise a wide range of residential, commercial and leisure facilities. For designing the new Thames-side, efforts have been made in balancing the needs of planning authorities and the perceptions of local residents with those of the developers, resulting generally in an exercise of creativity, negotiation and compromise.

It is said that riverfront developments have a certain romantic feature. The Thames has been a place of imagery and association for at least 2000 years. Some parts of the river are commonly associated with Royalty, some with famous writers such as Charles Dickens and some with Medieval London. The major ones bring to mind the heyday trading in the Pool of London and Docklands. Some recall the character of Victorian and Edwardian times with the building of the embankments north and south of the Thames in Central London. Designing the new developments, such associations could be taken into account for the particular part of the river and serious consideration must be made of how to accommodate the requirement for public access along the riverfront. Recent developments with Thames-side paths are London Bridge City at Southwark and Butlers Wharf in Bermondsey.

New Docklands Waterside

As you leave Bank station on the Docklands Light Railway (DLR), there is something about the jerky start of the hi-tech train that you are about to go on a journey into the future. Over the past three decades of the 20th century the landscapes of Docklands watersides have changed almost beyond recognition. As the train travels overland, you see the giant cranes, the converted warehouses and the glittering skyscrapers. Throughout the seventeen years of its existence, the London Docklands Development Corporation (demised in 1998), spent £2billion of government money and lured another £7billion of private capital into the area.

Canary Wharf is the epicentre of Docklands and dominated by that huge Christmas tree of a building, the tower at No.1 Canada Square. This tower is the symbol of Docklands and a beacon for London that signals to the business folk of the third millennium to go east. You walk into one of the glass-domed foyers and immediately know that it is not just an ordinary business place.

Thames Path & Dockside Access

For many years there has been concentrated effort to provide public access to the River Thames and dock edge. This has reinforced rights of access by providing pedestrian signage. It has also contributed to the Countryside Commission's Thames Path initiative by providing comprehensive signing of the Thames Path route through Docklands between Tower Bridge and Greenwich. The Thames Path runs between its source at Kemble in Gloucestershire and the Thames Barrier in Greenwich. Around 60km of cycle route have been implemented throughout the Docklands area; just over half of this is in the form of off-street routes making them both safe and attractive to use. Almost all of Docklands is now accessible by cycle routes making the area an important cycle-friendly part of London. (see maps opposite)

New Pedestrian Bridges

The need for new pedestrian bridges in Central London and throughout East London Docklands arose from the severance caused by past shipping along the Thames up to London Bridge by the large expanses of dock water. Ways had to be found to overcome these problems in a way that recognised the integrity of the river and dockscapes with sometimes the conflicting needs of water users, pedestrians and developers.

South Quay Footbridge - The S-shaped bridge spans 180m across south Dock near Canary Wharf. It is made up of two identical cable-stayed sections, one of which rotates to allow boats to pass through. Each half of the bridge features a 32m tall steel raking mast from which the deck is supported by cable stays. Bridge spotters are advised to see the structure at night, when the bridge is illuminated.

West India Quay Footbridge - The gently arched 90m pontoon bridge has a hollow stainless-steel spine beam spanning the main pontoon supports which are tied to the dock bed by anchor weights. A central opening section gives access to the west end of the dock.

Royal Victoria Dock Footbridge - This bridge, opened to the public in July 1999, provides a link route for developments on both sides of the dock and to the Thames riverside. Designed for a 120-year lifespan, its deck level has been set 15 metres above water level permitting sailing for most types of river boats below. The superstructure crosses the dock in a single span of 130 metres. The main structure is an inverted truss with six masts bracing cable stays and supporting the deck. The Docklands Light Railway Custom House Station is to the north of the bridge.

St Saviour's Footbridge spans the entry into the old dock of that name with surrounding new developments, connecting the riverside walkway just downstream of Tower Bridge at Butlers Wharf with the area of Bermondsey. The fine structure in gleaming steel was recently built at a cost of £600,000. By night the lines of the bridge are picked out by fibre-optic lighting. The bridge is located in the heart of an outstanding conservation area, next to converted Victorian warehouses and right on the Thames river front just downstream of Tower Bridge. The cable-stayed structure has a swinging central span of 14.5m and side spans are supported on timber trestles. The masts and cables reflect the nearby dockside cranes and the whole structure is light enough to be swung by hand.

Limekiln Footbridge - spans Limekiln Dock linking Limehouse with the new walkway past the giant Canary Wharf on the north bank of the Thames. The dock gets its name from the lime kilns that once stood here, producing lime for mortar and for a nearby porcelain factory. It has a single mast supporting the curved bridge deck, which rotates to open the way into the dock. There are uplighters and fibre-optics to give extra drama to the bridge at night. Walkers and cyclists using the Thames Path at Limehouse are familiar with this landmark. The 34m span cable-stayed structure is a counter weighted swing bridge. The deck which is curved in plan, is hung by six cables from a single mast located at the quayside. Two backstay tension members transfer the compression forces from the swing deck to a concrete-filled counterweight. The bridge is opened by slewing the deck, mast and counterweight using two electric motors.

The Millennium Footbridge - The 332 metre long crossing connects the New Globe Theatre and Bankside Gallery to St Paul's Cathedral on the north bank. It is a metal frame of tension and compression members structure similar to a cradle. Each end of the bridge's deck is hung by suspension cables. The mid section rests on struts which stand on the cables as they descend below deck level.

Unfortunately, as with every design which pushes to the edge of current knowledge, the bridge suffered from some unfortunate dynamically induced vibration during the first days of its opening. Though these vibrations were uncomfortable they presented no immediate danger to the public. However, in order to ensure the bridge works to its optimum capacity, researchers and designers are investigating this unique problem in order to enhance the bridge's performance under working conditions.

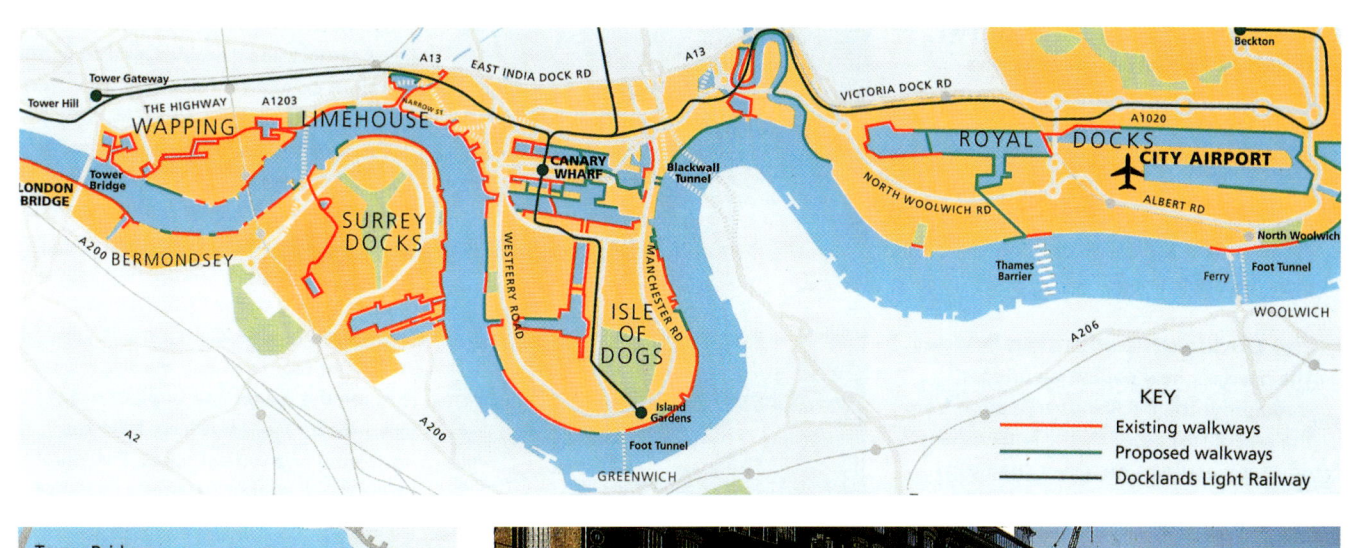
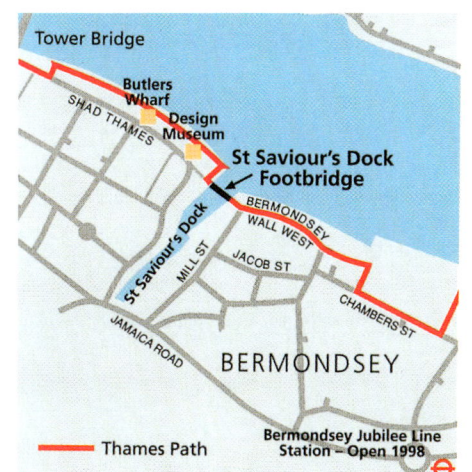

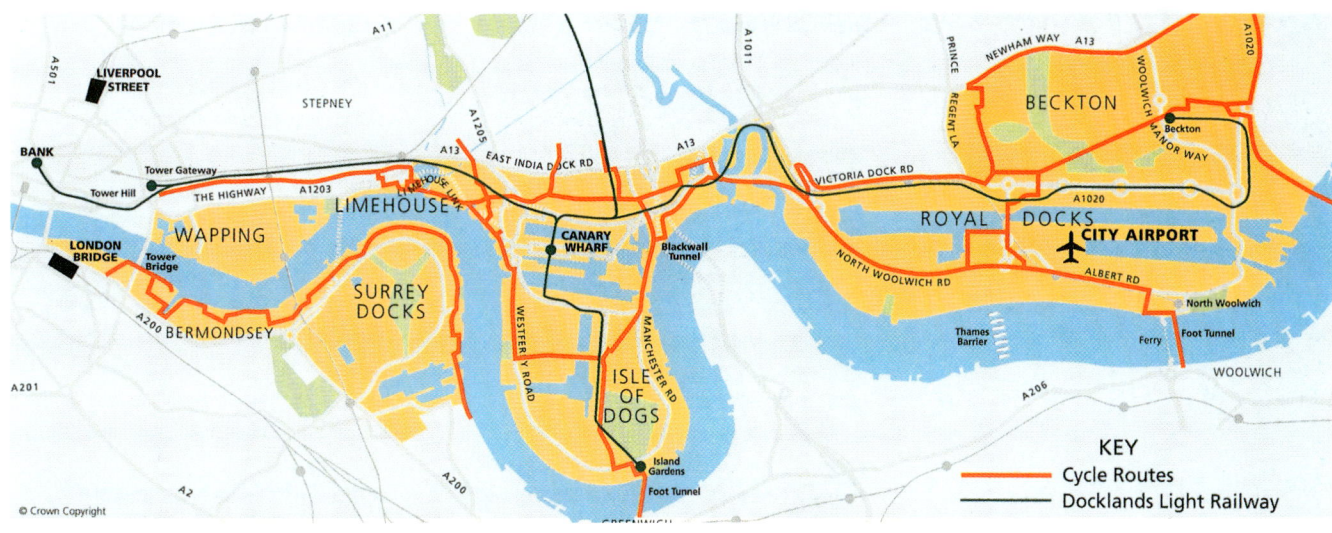

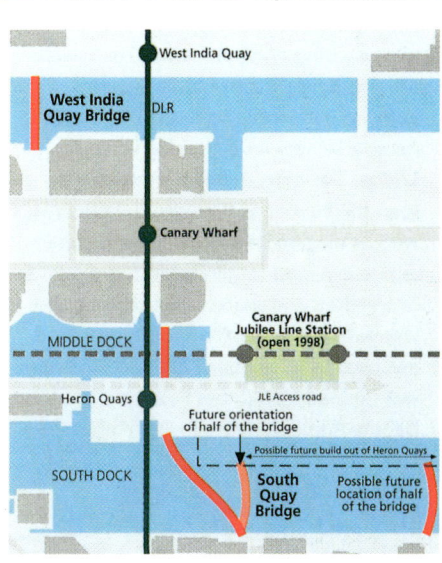

Thames Wildlife

Pollution History

Today the naturalist can find plenty of wildlife along the River Thames and on its adjacent marshes. A wide range of bird species winter along the river and many of them remain during the summer to breed. However, it was not like this during the last century. Until 1970 the Thames was rather dirty and polluted as other rivers in industrial cities of Europe. Severe problems existed in the first half of the 19th century but considerable improvements were made when new sewage works were developed at Beckton and Crossness. But the river still had outfalls discharging untreated sewage.

It took about two months for the pollutants to reach the North Sea, 64 km from London Bridge. The river had little oxygen but plenty of rubbish. During the two World Wars, matters got worse because of the damage to sewage plants.

In the 1950s a determined effort was made to improve the quality of the river. Beckton and Crossness sewage works were modernised at considerable expense. The effluent pumped into the river became as pure as possible. Industrial effluent was also treated by manufacturing factories, with hard detergents excluded. Around 1963 it was observed that fish had started to return upriver. Until then only eels had survived because they breathe oxygen from above the water's surface. With the closure of the docks the pollution from river traffic was further reduced. By 1975 over eighty species of freshwater and sea fish were identified along the tidal reach of the river up to Teddington and birdwatchers reported wintering flocks of pochard along the Thames, particularly around Rainham, Barking and Dartford. Soon these were joined by mullards, tufted ducks, teals, shoeveler and pintail. Several thousand ducks live at Thamesmead and Silvertown, near the Thames Barrier.

Thames Marshes and Islands

The ducks, Canada Geese and White Swans which winter along the Thames are attracted by the riverside marshes. This habitat is unfortunately being threatened by modern developments. In the early 1930s there were nearly 7000 acres of Thameside marshland but this was reduced to about 2000 acres in 1980, a decline of over 70 per cent. Over the past two decades efforts have been made to identify these sites and protect them.

Between Teddington and Chiswick in West London, there are several wooded islands called aits or eyots, which provide winter homes for cormorants and roosting sites for herons. Birds seen include the great crested grebe, dabchick and kingfisher. Nearby Chiswick is Syon Park which is an excellent site of natural riverbank characterised by inlets and swamp cypress.

However, the best area for birds are the ancient Rainham Marshes in Essex, where there are grazing marshes and man-made lagoons. The Scandinavian teal and Russian pintail are regular visitors to these marshes. Other birds include the reed warbler, wagtails and redshank.

The majority of London's riverside marshes have been lost in a century but Londoners must endeavour to ensure that future generations will be able to enjoy a beautiful wildlife environment of the fragments that are still intact.

The illustrations below show pictures of wildlife along Thameside with Glover Ait in the river.

London Wildlife Trust

London contains a large number of important wild sites. London Wildlife Trust was founded in 1981 to protect these habitats and manage over sixty of the most precious ones including the Isleworth Ait, a small island sanctuary in the Thames. Thamesmead is an excellent reserve south of the river, containing woods, wetlands and meadows; habitats which support over 200 plants and 100 birds.

The Trust actively campaigns to protect threatened sites and thanks to their great efforts, along with English Heritage, 170 species of birds have been protected on Rainham, Wennington and Aveley marshes on the east side of London.

Thames Bridges Heritage Trail

River Thames Crossings

Throughout the 17th and 18th centuries the River Thames was a source of water and the main avenue of communication for the Capital. Gradually civil engineers and architects built bridges, pumping stations, locks and docks, therefore beginning to control and alter the character of the great river. The finest views of the Capital remained and were an inspiration for international artists and local engravers. They created paintings showing a bustling waterfront of sailing ships, barges and ferries against a background of churches, warehouses and crowded quaysides. The City of London was a vibrant place, where life and work provided an infinite source for famous painters of the period. London museums and public institutions possess magnificent collections of these artists' works. At the beginning of the 18th century London Bridge was still the only crossing on the river, but by the mid 19th century there were ten bridges built by civil engineers. Today there are 20 bridges and an historic trail is briefly described below starting upstream.

Hampton Court Bridge - This bridge, the third to stand on the same site, is constructed of reinforced concrete and Portland stone. It opened in 1933, the farthest crossing in London.

Kingston Bridge - Built 1825-28, this bridge has fine rusticated Portland stone arches with bold cornice.

Richmond Bridge - One of the first bridges in London it was built 1774-77. There are fine arches in Portland stone.

Twickenham Bridge - Built in 1933 this arch type bridge of reinforced concrete, which has bush hammered and bushed concrete surface finish, is the first bridge at this point on the river. Ferries crossed over here as long ago as 1659.

Kew Bridge - Formerly called Edward VII bridge the present bridge was built of Scottish granite and opened in 1903.

Chiswick Bridge - Designed by Dryland and Baker, the bridge opened in 1933. It has three arches of reinforced concrete faced with Portland stone.

Hammersmith Bridge - In 1827 the first Hammersmith suspension bridge was opened. Being of narrow construction, 14 ft. (4.2m), it was inadequate to take later traffic and eventually became unsafe. The present suspension bridge was completed in 1887. This bridge in turn has recently been strengthened to overcome the problem of strain through traffic load.

Putney Bridge - Designed by Sir Joseph Bazalgette and constructed 1882-86 of granite with five segmental arches and rusticated voussoirs, its handsome cast iron lamp standards surmount the plain parapet in the centre of each arch.

Wandsworth Bridge - This modern bridge of 1940 is constructed of huge steel cantilever beams pinned down on the banks and resting on two narrow streamlined piers, widely spaced.. The engineer was Sir Pierson Frank and the architects were E Wheeler and F Hirons. The previous bridge built in the 1870s was a continuous lattice girder type but by 1926 extensive repairs were necessary.

Battersea Bridge - The present bridge, opened in 1890, was designed by Sir Joseph Bazalgette and is made of wrought iron and steel.

Albert Bridge - One of the few remaining original bridges designed by R N Ordish which was opened to the public in 1873. In 1884 the ropes became dangerous and were replaced by steel chains. In 1970 the weight limit was reduced from 5 to 2 tons and in 1972 the bridge was closed so that considerable strengthening could take place including a new lighter bridge deck and replacement of rusted cast iron with steel plates.

Chelsea Bridge - This modern steel suspension bridge of 1937 replaced an older iron suspension bridge. The architects were Topham Forrest and E P Wheeler, the design was by the Consulting Engineers Rendell, Palmer and Tritton. The old bridge of 1858 was a narrow and ornate wrought iron suspension bridge first named Victoria Bridge. The new structure by comparison is almost free of ornamentation.

Vauxhall Bridge - The present bridge was opened in 1906. It has a steel superstructure and the abutments are faced with granite. The old bridge was the first iron bridge across the Thames built in 1816. A temporary bridge was provided in 1898 to serve until the completion of the present structure.

Lambeth Bridge - A suspension bridge once stood slightly downstream of the present bridge. It was demolished in 1929 and the new bridge opened in 1932. It is an arch structure of steel with circular arch ribs.

Westminster Bridge - A fine Victorian bridge built 1854-56 to the design of engineer Thomas Page in harmony with its historic surroundings. The seven Gothic style cast-iron arches rest on piers and abutments of grey granite. The handsome balustrades, lamp standards and spandrels match the Houses of Parliament on the north side of the bridge. Its predecessor, built between 1739-1750, was the one immortalised by William Wordsworth in 1803 in his sonnet, inspired by the view from the bridge at dawn.

Waterloo Bridge - The present bridge was completed in 1944 replacing John Rennie's bridge opened in 1817 which was, at first, known as Strand Bridge.

Blackfriars Bridge - Built 1860-69 by Cubitt & H Carr, the iron bridge has five arches resting on granite piers. Queen Victoria replaced an earlier structure of 1768. Some of the references refer to the earlier bridge.

Southwark Bridge - The old Southwark Bridge designed by John Rennie and finished in 1819 was rebuilt and reopened in 1921. The new structure was much wider with improved gradients.

London Bridge - The elegant new London Bridge, completed in 1972, has a wealth of history to live up to in its predecessors. Old London Bridge existed for about 600 years and was the first stone bridge over the river. Started in 1176 it saw and felt aspects of London life through until 1831. The second London Bridge designed by John Rennie was a masonry structure which became inadequate for modern needs when widening became impossible due to structural faults. The stonework however now stands at Lake Havasu City in Arizona, sold to the Americans for one million pounds.

Tower Bridge - Perhaps the best known bridge in the world this was started in 1886 and opened in 1894. It is a huge piece of mechanism constructed of steel, encased in towers built of highly ornate masonry. It has three spans, the middle of which is a double-leaf bascule operated by hydraulic machinery. The machinery has never failed but the bridge rarely opens now due to the decline in the use of the docks. It remains a monumental landmark dominating the river landscape.

Illustrations

The illustrations on the next two pages give details of present and past bridges and some landmarks along the River Thames.

Peaceful scene and fishing near Richmond Bridge, c1880s, a splendid place to relax today.

The Albert Bridge and Chelsea Embankment, c1880s, the bridge has suffered from structural problems.

Rowing near the old Hammersmith Bridge, c1880s, boathouses still exist along the riverside.

Vauxhall Bridge from Nine Elms Pier on south bank, c1880s, where New Covent Garden Market was built.

The Medieval Lambeth Palace and River Pier on south bank, c1880s. The Albert Embankment was later built.

Blackfriars Bridge, sailing barges and boats looking north towards St Paul's, c1880s.

The Royal Hospital and Thames sailing barges at Greenwich, c1886. The area has one of the finest views in London.

The old London Bridge looking north towards the City and the Monument, c1880s.

Hammersmith Bridge - Built 1887 on the piers of the original bridge of 1827.

This suspension bridge has iron cased towers with monumental anchorages.

Putney Bridge - Built 1886, the bridge has handsome cast iron lamps.

Wandsworth Bridge - A handsome modern bridge of 1940 with huge cantilever beams.

Battersea Bridge - A well proportioned structure of 1890 decorated with oriental motifs.

Albert Bridge - This suspension bridge, built 1877, has cast iron arches.

Vauxhall Bridge - Opened in 1906, this steel bridge replaced the first iron bridge of 1816.

Lambeth Bridge - A modern bridge of 1932 with steel arches bearing on granite piers.

Westminster Bridge - completed 1856 to match the Gothic style of the Houses of Parliament.

Waterloo Bridge - A modern concrete bridge built 1942 and faced with Portland stone.

Blackfriars Bridge - This iron bridge was opened by Queen Victoria in 1869.

Southwark Bridge - Opened by King George V in 1921, replacing an older one of 1819.

London Bridge - The concrete bridge constructed in 1972 has a history of 2000 years.

Tower Bridge - Since 1894 the bridge has been a symbol of London and a world-famous landmark.

Enjoy its museum and a panorama of London and the Thames from its high walkway.

London Safe from the Sea

Barrier Holds Back the Sea

In 1984 a possible catastrophe to London may well have been averted. The high tidal waters waiting to drown the Capital were officially tamed when the Queen declared the Thames Barrier open in May. The world's largest moveable flood barrier, which straddles the River Thames at North Woolwich, came into operation just in time. For the Capital is sinking at a rate of 0.3metre each century and the sea tides are rising higher.

Fourteen people were drowned when the river burst its banks in 1928. Three hundred were killed in floods on the east coast in 1953, when London came within 80mm of flooding. In 1978 the river came to within just 0.3 metre of breaching the city's defence walls.

Over one million people live in the flood plain of the Thames on the east side and centre of London. If the big flood had come it would have had considerable economic effects for the whole of the United Kingdom. The death toll could have reached many thousands, without the barrier protecting the city. Around one quarter of a million homes, offices and factories would have been hit by the deluge.

The damage could have run into billions of pounds and taken up to two years to rectify. It was in 1973 that the Government gave the go-ahead for a rising sector gate type barrier and after a series of difficulties, and at a cost of £450 million, Old Father Thames was finally under man's control - we hope!

Reasons for Flood Threats

The Thames Barrier was built to prevent a North Sea flood tide rushing up the river and devastating more than a third of the capital. This threat arises because Britain is said to be slowly tilting, causing London and the south-east to dip south eastward. It is claimed, from this theory, that at the end of the last Ice Age, the crust under Northern Scotland was compressed by very thick ice. As the ice melted it caused the south of Britain to move downward. Recent high tides have been measured more than 0.6m higher at London Bridge than a century ago.

A third factor is the formation of storm surge tides. High winds and bad weather conditions in the North Atlantic and the North Sea could cause high waves and surges travelling above sea level thus raising the Thames by about 2m, well above existing flood defences. The surge, a wall of water travelling at high speed above sea level, is forced down the wedge-shaped part of the North Sea to the bottle-neck of South East England. There it is funnelled up the Thames Estuary, growing higher as the river narrows upstream. If this surge came on top of an already high tide, without the Barrier, London would be flooded.

The gigantic barrier consists of a set of piers, between which are huge steel gates. During a flood alert the gates are lifted hydraulically to stop the incoming surge, and Londoners can stay safely in their beds!

Thames Barrier Rising Gates

The basic design requirements for the Thames Barrier scheme were to prevent surge tides passing upstream of its position, for it to be completely reliable in operation, to enable ships to pass safely and to cause minimum interference with the flow of the river. It consists of four main navigation openings of 61m clear span each containing a rising sector gate to give a channel depth of just over 9 metres, two smaller openings of 31.5 metres span with sector gates and four other smaller ones with conventional falling radial gates.

The sector gates are steel box girders whose cross-section is a segment of a circle. These are supported on both ends by double skin steel discs which are centrally pivoted about stub shafts projecting from the piers. The gates normally lie in a bed of the river in the scallop of a concrete cill. Each gate is operated by means of electrically driven hydraulic machinery. In operation, the steel discs, or gate arms rotate through approximately 90°, to take the gates up to their high flood prevention level. The discs can also be rotated through a further 90° to move the gates to their maintenance position above the river water level. The housing for the operating machinery is roofed with shiny stainless steel. The total weight of steel in the barrier is 51,000 tonnes and total volume of concrete used was 214,000 cubic metres.

Operation & Responsibility

Normally the barrier can go to a position of full closure in approximately 15 minutes, though with modes of operation involving partial closure this period can be longer. More time is needed however, for preparatory work before closure commences. Closures with water shooting under the gates form an important part of the emergency operation technique. With the data links wired into the central control room, command of the barrier has been entirely self-contained. Standard routine for operation is to check functions of all six rising sectors first. Vigilance is concentrated on the multiple systems guarding against electrical or mechanical failures; so far they have not occurred.

The barrier has proved an enormous tourist attraction as a dramatic piece of engineering close to the heart of a great city. The barrier's political worth was never missed by the former Greater London Council (GLC) which was charged with overseeing the project. When the GLC was disbanded in 1986, it was transferred to the Thames Water Authority with the barrier becoming its most prized possession.

Barrier Point, Royal Docks, luxury apartments facing the Thames Barrier, c2000.

Thames Barrier

Giant steel links and beams work the sector gates. Inset: A memorial to the first settlers who sailed to America.

A walkway along the south bank of the river with one of the gates nearest the railings closed.

The diagrams show the formation and travel of possible surges and waves into the Thames Estuary, resulting in the Barrier's use.

This photograph shows the giant sector gates raised above water level for maintenance.

There are four 61m gate openings to allow shipping through. The picture shows a ship being guided by tugs.

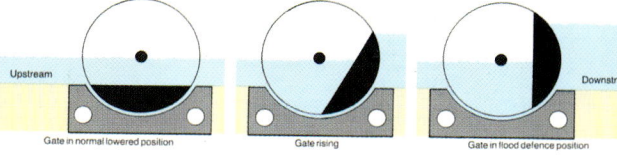

These illustrations show the operation of the gates during flood surge when the gates are fully raised as shown in the top picture.

Acknowledgements

I would like to thank sincerely Linda Day for her help, excellent typing and patience in preparation of the whole manuscript with great care. I am also very grateful to Joyce O'Neill for typing, loyalty and continued help. I am grateful to Dr Paul Smith for advice and to Terence O'Connell for general assistance. Special thanks are due to Tom Juffs for his enthusiasm and unstinting support throughout the project. I would like to express my gratitude to Dr John Grubert for proof reading and assistance with the layout of illustrations. Thanks are due to Dave Hobson and his staff of Lipscomb Printers and to Steve Cook at Stadium Graphics. I am most grateful to my wife, Irene, for her support and patience over many years. I would like to express my thanks to my institution, the University of East London, for its support of the research work.

I am deeply indebted to previous writers, individuals, photographers, estate agents and many organisations whose information helped greatly with the preparation of the book. These include Thames Water, Port of London Authority, former London Docklands Development Corporation, Department of Transport, New Civil Engineer, Docklands News, Meridian Magazine, Docklands and City Magazine, Chorley & Handford and Museum of London.

To members of the public, visitors, students, teachers and scholars world-wide, who have kindly supported our twelve book publications over the past two decades, some of which are in their seventh and eighth editions, I express my deepest appreciation. The books are standards and are providing an essential public service in Great Britain.

Inside Back Cover Illustrations

Docklands Waterside Developments

1. Tesco Shopping Precinct overlooking Canada Water in Surrey Quays.
2. New housing along the picturesque Wapping Canal.
3. Waterside housing and sailing at Greenland Dock.
4. Canary Wharf, the new financial centre of London.
5. New office complex overlooking the old East India Dock.
6. The beautiful Hays Galleria precinct with shops and cafés overlooking the Thames.
7. Historic Greenwich with its Cutty Sark and foot tunnel entrance.

Information

A unique set of nine internationally acknowledged research books have been published on the history, heritage, regeneration, infrastructure, walks and millennium of London and Docklands. They are ideal for teaching and research in schools and colleges as well as for libraries, visitors, the general public and walkers.

"London Water Heritage" Portrait in Words and Pictures (ISBN 1-8745-36-40-6),
"London and Docklands Walks" The Explorer (ISBN 1-8745 36-25-2)
"London Docklands" Past, Present and Future (ISBN 0-091987-81-6),
"Discover London Docklands" A-Z Illustrated Guide to Modern Docklands
(ISBN 1-874536-00-7),
"European Docklands", Past, Present and Future (ISBN 0-901987-82-4),
"London Illustrated" Historical, Current and Future (ISBN 1-874536-01-5),
"London Docklands Guide" Heritage and Millennium Exhibition
(ISBN 1-874536-03-1).
"London Millennium Guide" Education, Entertainment and Aspirations
(ISBN 1-8745-36-20-1).
"Dockland" Historical Survey
(ISBN 0-901987-8).

For further information visit the University of East London Web Site on
www.uel.ac.uk
or telephone 020 8223 2531,
Fax 020 8223 2963.

**PLEASE ORDER THROUGH:
RESEARCH BOOKS,
P O. BOX 82,
ROMFORD, ESSEX,
RM6 5BY, GREAT BRITAIN**

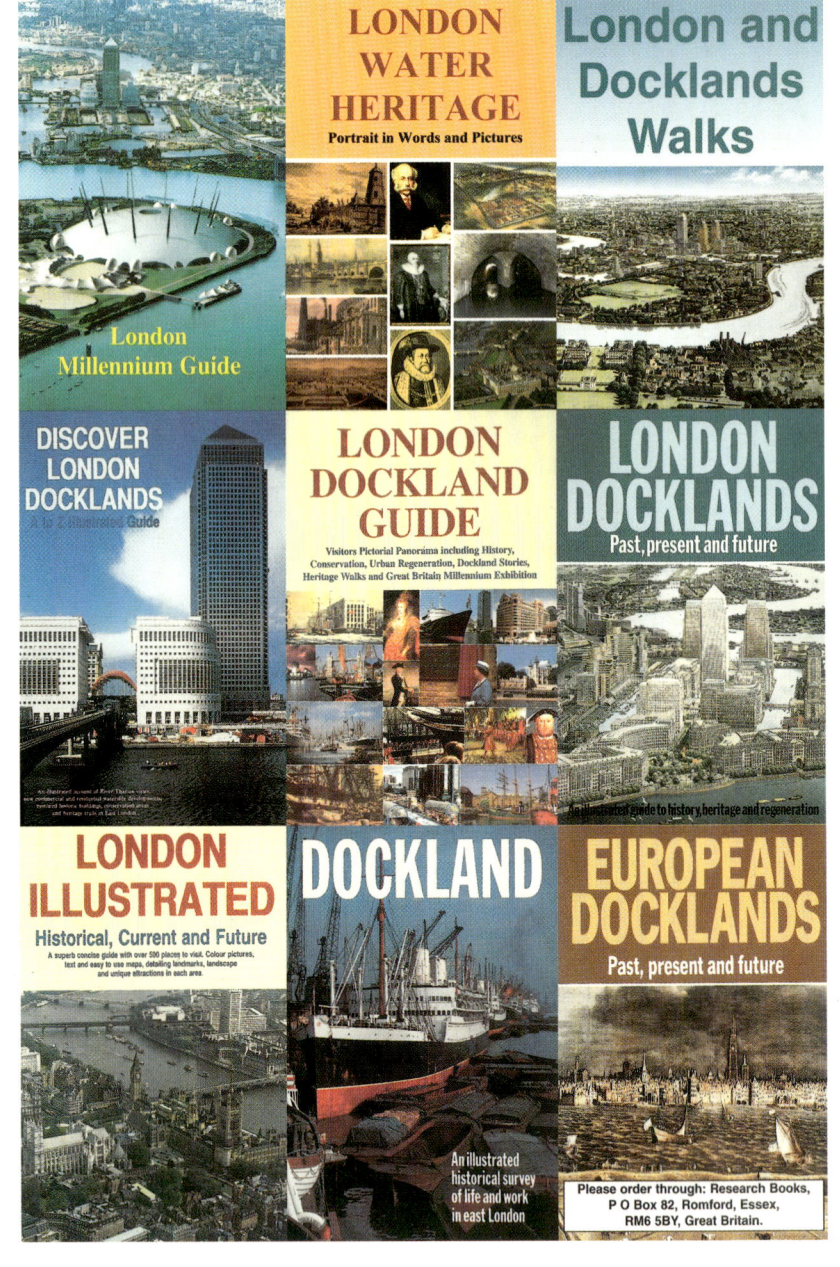